EVALUATION OF RAIL TECHNOLOGY

Human Factors in Road and Rail Transport

Series Editors

Dr Lisa Dorn
*Director of the Driving Research Group, Department of Human Factors,
Cranfield University*

Dr Gerald Matthews
Professor of Psychology at the University of Cincinnati

Dr Ian Glendon
*Associate Professor of Psychology at Griffith University, Queensland,
and President of the Division of Traffic and Transportation Psychology
of the International Association of Applied Psychology*

Today's society confronts major land transport problems. Human and financial costs of road vehicle crashes and rail incidents are increasing, with road vehicle crashes predicted to become the third largest cause of death and injury globally by 2020. Several social trends pose threats to safety, including increasing vehicle ownership and traffic congestion, advancing technological complexity at the human-vehicle interface, population ageing in the developed world, and ever greater numbers of younger vehicle drivers in the developing world.

Ashgate's Human Factors in Road and Rail Transport series makes a timely contribution to these issues by focusing on human and organisational aspects of road and rail safety. The series responds to increasing demands for safe, efficient, economical and environmentally-friendly land-based transport. It does this by reporting on state-of-the-art science that may be applied to reduce vehicle collisions and improve vehicle usability as well as enhancing driver wellbeing and satisfaction. It achieves this by disseminating new theoretical and empirical research generated by specialists in the behavioural and allied disciplines, including traffic and transportation psychology, human factors and ergonomics.

The series addresses such topics as driver behaviour and training, in-vehicle technology, driver health and driver assessment. Specially commissioned works from internationally recognised experts provide authoritative accounts of leading approaches to real-world problems in this important field.

Evaluation of Rail Technology
A Practical Human Factors Guide

EDITED BY

CHRIS BEARMAN
Central Queensland University, Australia

ANJUM NAWEED
Central Queensland University, Australia

JILLIAN DORRIAN
University of South Australia

JANETTE ROSE
University of South Australia

DREW DAWSON
Central Queensland University, Australia

CRC Press
Taylor & Francis Group
Boca Raton London New York

CRC Press is an imprint of the
Taylor & Francis Group, an **informa** business

CRC Press
Taylor & Francis Group
6000 Broken Sound Parkway NW, Suite 300
Boca Raton, FL 33487-2742

First issued in paperback 2017

© 2013 by Chris Bearman, Anjum Naweed, Jillian Dorrian, Janette Rose and Drew Dawson and the contributors
CRC Press is an imprint of Taylor & Francis Group, an Informa business

No claim to original U.S. Government works

Version Date: 20160226

ISBN 13: 978-1-4094-4243-1 (hbk)
ISBN 13: 978-1-138-07420-0 (pbk)

Visit the Taylor & Francis Web site at
http://www.taylorandfrancis.com

and the CRC Press Web site at
http://www.crcpress.com

Contents

List of Figures

List of Tables

About the Editors

Chris Bearman is a Senior Research Fellow at the Appleton Institute of Central Queensland University. His research involves conducting industry-focused studies in the areas of occupational health and safety, human factors and applied cognitive psychology. He works closely with industry partners and government organisations around the world to produce research that has both a strong theoretical underpinning and a robust application to industry. In relation to technology human factors, Chris has worked with the Federal Aviation Authority and NASA (National Aeronautics and Space Administration) on projects examining the implementation of new technology into the US airspace system. He has also worked with numerous rail organisations to examine aspects of designing, implementing and evaluating in-cab and train-control technologies. In other work, he has examined the pressures that lead pilots to make poor decisions; how coordinated decision-making breaks down in teams; the role of the supervisor in creating workplace safety; and how people use prior experience to solve novel problems.

Anjum Naweed is a Senior Postdoctoral Research Fellow at the Appleton Institute of Central Queensland University. He has experience with rail human factors that spans over a decade. In 2001 he validated a proprietary car simulation game for psychological research as part of his Honours project. His Master's project in 2004 extended this work to explore the influence of auditory feedback on speed choices. In 2005 he spent a year as a research scientist investigating alarm and alerts confusability in train cabs. He obtained his doctorate in 2011 after developing a laboratory train simulation paradigm and investigating decision-support and information design in train displays. In 2012 he was awarded the Body of Knowledge prize at the Asia-Pacific Simulation Training and Technology conference for best contribution to standards development, capture of best practice or contribution to the simulation community. Anjum currently leads six research projects with the Australian Cooperative Research Centre for Rail Innovation, investigating the area of safety and security on topics involving train driver psychology, complex decision-making, knowledge representation, level crossing design, work health and safety, and participative processes at work. He has experience with a wide range of qualitative and quantitative research methods, and is passionate about exploring all aspects of human factors, workplace culture, and the relationship between humans and machines.

Jillian Dorrian is a Senior Lecturer in Psychology in the School of Psychology, Social Work and Social Policy, at the University of South Australia. Since graduating with her PhD in 2003, she has co-authored more than 60 articles and book chapters. Her work has been cited more than 500 times in the scientific literature. Jillian specialises in the area of human sleep, biological rhythms and performance. She has research experience in both the controlled laboratory environment and in the field, and has worked extensively with the Rail and Healthcare Industries investigating sleep loss, workload, operational performance, safety and health. She worked closely with the Australian Shiftwork and Workload Consortium (1995–2005), a collaboration of industry, unions and science. As part of this, Jillian has synergised the use of rail simulators and locomotive-based data-loggers to pursue rail research. She has focused on the use of these complementary research tools to understand issues related to safety and performance in train driving, particularly fatigue, self-awareness and cognitive disengagement. She currently collaborates with the Australian Cooperative Research Centre for Rail Innovation on a number of applied simulator research projects looking at knowledge representation, workplace culture, and generating evidence-based guidelines for single and dual-driver operations. Jillian also loves analysing large, complex datasets, gaining her Master of Biostatistics in 2012.

Janette Rose is a PhD student at the University of South Australia, conducting research into the human factors implications of introducing new technology into the train cab. Her particular interests are in the development of improved tools to measure subjective situation awareness, and user resistance to technology. In the course of her research, Janette has worked with drivers and managers from several rail organisations around Australia. Janette attained her Honours degree in Psychology in 2008 and became a PhD student with the Human Factors Group at the University of South Australia in 2009. Her interest in rail human factors was piqued during research for her Honours thesis and her PhD follows on from this initial research. She has published a paper on the use of task analysis to evaluate the effects of information support technology on the situation awareness of train drivers. She has also contributed papers to conferences on a new subjective measure of situation awareness that she developed during the course of her PhD research.

Drew Dawson is the Director of Central Queensland University's Appleton Institute and the university's inaugural Engaged Research Chair. He has been involved in Australian Rail Research for a number of years and in 2008 was appointed as the Program Leader in CRC for Rail Innovation. He is directly involved in four projects in relation to this, but as director of the institute oversees all 12 projects related to the CRC for Rail Innovation. Drew is an established international researcher and has published both nationally and internationally (more than 150 C1 DEST publications in international peer-reviewed journals) with a Web-of-Science career citation count around 1400. He has been instrumental in championing and establishing significant research collaborations between industry, regulatory and university sectors for projects relating to safety management and workload in rail and aviation sectors in Australia, North America and Europe. Drew consistently works as an expert witness in national and international rail safety cases.

List of Contributors

Ganesh Balakrishnan, Central Queensland University, Appleton Institute, Adelaide, Australia

Chris Bearman, Central Queensland University, Appleton Institute, Adelaide, Australia

Verna Blewett, Central Queensland University, Appleton Institute, Adelaide, Australia

Drew Dawson, Central Queensland University, Appleton Institute, Adelaide, Australia

Jillian Dorrian, School of Psychology, Social Work and Social Policy, University of South Australia, Adelaide, Australia

Gareth Hughes, Railcorp NSW, Sydney, Australia

Airdrie Long, Considered Solutions Pty Ltd, Sydney, Australia

Anne Maddock, Railcorp NSW, Sydney, Australia

Anjum Naweed, Central Queensland University, Appleton Institute, Adelaide, Australia

Janette Rose, University of South Australia, Adelaide, Australia

Andrea Shaw, Shaw Idea Pty Ltd, Mount Egerton, Victoria, Australia

Kirrilly Thompson, Central Queensland University, Appleton Institute, Adelaide, Australia

Foreword

Over the past decade, the Australian rail industry has witnessed an overwhelming flow of new technologies, all seeking to improve the safety and efficiency of modern rail operations. Over the next decade, we will undoubtedly see a further acceleration of the adoption of new rail technologies. While the introduction of new technologies holds a great deal of promise for the rail industry it is important that decisions about investment in new technologies are based on a solid base of evidence about their limitations and benefits.

This book is a practical guide to evaluating the human factors issues of technologies in the rail industry, and represents a tool-kit of methods that provides resources for people at different levels in the process of making decisions about that technology. The book is designed to assist those making decisions about investing in rail technology, those tasked with developing and implementing a technology and the human factors consultant who may be contracted to assist in the evaluation. As such this work helps plug a gap in respect of published material and standard methodologies to address the human factors implications of new technology in the rail context.

This book is not only a comprehensive overview of how one might think through the issues associated with introducing a new technology, so that the likelihood of error and unintended consequences are reduced, but it also illustrates the complexity of issues around successfully introducing new technology.

The editors of this book have led the way in finding innovative solutions to rail human factors problems, particularly those surrounding the introduction of technology. This book presents their expertise in this area together with valuable contributions from their colleagues about particular issues and methods of evaluating technologies. This book is an excellent example of collaborative research delivering world class knowledge and information to the rail industry.

The Cooperative Research Centre (CRC) for Rail Innovation has been an active participant in research in this area and I'm delighted that we have been able to finance and sponsor this significant and innovative work. Australia is 'punching above its weight' in the development of new technologies in the rail industry and having access to this world leading tool-kit is a crucial part of managing the risk and successfully introducing new technologies that will make the rail industry safer and more efficient.

David George
CEO
Cooperative Research Centre (CRC) for Rail Innovation, Australia

Acknowledgements

The authors and editors of this book are grateful to Janette Rose for her technical editing. Her attention to detail and dedication to the task has been a great asset. We would also like to thank: Judith Timoney for her editing assistance; Adrian Hurley and Gareth Hughes for their assistance in shaping the overall content of the book and for their helpful advice from an end-user perspective; David George, Chris Gourlay, Kellie Dyer and the rest of the team at the CRC for Rail Innovation for their enthusiastic support from the initial concept of the book to its final production; and Matthew Thomas and Benjamin Brooks for their helpful advice and comments on early conceptualisations of the book contents.

Chapter 1

Introduction:
The Promise and Perils
of New Technology

Drew Dawson

Central Queensland University, Appleton Institute, Adelaide, Australia

Chris Bearman

Central Queensland University, Appleton Institute, Adelaide, Australia

Anjum Naweed

Central Queensland University, Appleton Institute, Adelaide, Australia

Gareth Hughes

Railcorp NSW, Sydney, Australia

In the last decade the rail industry has witnessed an unprecedented flow of new technologies predicated on improving the safety and efficiency of modern rail operations. Over the next decade we will see an even greater penetration of these twenty-first-century technologies into an industry where lower levels of technology are still common. The introduction of new technology in any industry or workplace is a complex socio-technical phenomenon and it carries the potential for significant improvement with a justifiable concern about how the technology might also change the nature of work in unpredictable ways. This is especially the case in the rail industry where the historically dominant technologies have been mechanical rather than digital and the dominant organisational cultures are often relatively resistant to change.

This book is directed toward those in the rail industry who are charged with introducing new technology and ensuring that we do not fall foul of the law of inadvertent consequence. To that end, it is directed toward three types of readers. The first group is rail managers contemplating the introduction of new rail technologies. In deciding whether to introduce a new technology, this group typically needs to determine its cost-benefit analysis and ensure that the technology is implemented in a way that delivers the promised benefits. These readers will require high-level conceptual advice on the kinds of issues that need to be considered when contemplating the introduction of new technology. They will also need to be able to define the key issues associated with the design, implementation and evaluation of new technologies, but given their high-level vantage point, they will be less concerned with the details of this process.

The second group of readers is those charged with the implementation of a new technology into a specific rail operations setting. These readers will need to have a more detailed understanding of the specific technology being proposed and will be intimately involved in its design, implementation and evaluation. They will need specific advice on how to undertake each of these phases of the process and gain access to best-practice approaches. They will also need to provide operational answers to the questions raised by rail managers and ensure that the introduction of any new technology follows a sound process. This group of readers is, in effect, responsible for ensuring that an organisation has exercised due diligence in the selection and introduction of a new technology.

The third group of readers is human factors professionals who may be asked to provide independent advice or work on the introduction of new technologies but lack specific experience in a particular technology or in the rail industry in general. This book provides that group of readers with a number of new tools, some specific case studies of evaluating rail technology, and examples of how general theoretical principles and practice in Human Factors have been operationalised in a rail setting.

Each of these reader groups have different needs and, as a consequence, may well read this book in different ways. For the manager deciding on whether to investigate the feasibility of a new technology, the more generalist chapters around technology introduction (such as chapters 2, 3 and 7) will be of the most benefit initially. These will present the reader with key technology-related human factors issues for introducing technology, describe how the associated tasks may be evaluated, and demonstrate how technology may also be resisted. Similarly, the case studies that are presented in the specified chapters provide both exemplars and cautionary tales around ensuring that new technologies deliver on their technical promise. The two train simulation chapters (chapters 8 and 9) illustrate the seductive nature of technology and demonstrate how the purpose of technology may become easily disconnected from its goals during the early specification stages. For those from a primarily technical background (possibly with limited experience of the social and psychological issues that can arise around the introduction of new technologies), the case studies and chapters 2 and 3 will be particularly instructive.

For those in the second group of readers, the methodological chapters (chapters 4 to 10) will be of primary relevance. In these chapters, a sound understanding of the methodological challenges around participatory design and how to consult effectively is presented. Readers can learn about mock-ups and their surprising utility (chapter 4), the ways in which technology can be evaluated by structured qualitative research methods (chapter 5) and how a consideration of a technology's past, present and future can reveal striking insights. Understanding the methodological rigour required for effective consultation, participatory design, monitoring and evaluation of new technologies will be critical for those charged with ensuring that an organisation effectively fulfils its legal obligation to exercise due diligence around the introduction of a new technology. In many cases, the introduction of a new technology in organisations characterised by 'high engineering' cultures can be frustrated by the failure to understand the social, psychological or even industrial consequences of a new technology. The 'Luddites' of the nineteenth century still persist within many organisations and there is a clear need to sell the case for change to those required to use, adapt to or be displaced by a new technology. These issues are discussed in the chapter on resistance to technology (chapter 3).

The third group of readers, the human factors professionals looking to acquire specific knowledge and/or expertise in an operational rail setting, will likely find both the theoretical and methodological aspects of the chapters of considerable interest. Chapters 5, 8, 9 and 10 in particular provide discussions about theoretical issues concerning their subject matter. There are an increasing number of human factors professionals entering the rail industry to meet demand created by an increased focus on safety and a growing awareness of the need for human factors expertise. As a consequence, human factors professionals entering this industry may want to understand the peculiarities of the rail domain, and the ways in which general principles and practices have been specifically contextualised and/ or operationalised by the rail industry. Therefore, the chapter that describes task analysis (7), the two chapters (8, 9) exploring train simulation, and the last chapter (10), which investigates how situation awareness may be measured in rail, will be of particular interest to this group of readers.

Given the diversity of the readership, the book is, by definition, the type of publication into which one can *dip into* in many ways. It may even be the case that different chapters will be more or less relevant at different points in the process of introducing new technologies to an organisation. On the one hand, this book is a tool-kit, and different readers may use the perspectives contained in these chapters on a very much 'as needs' basis, but on the other hand, it contains a number of integrating themes that will reward those who read it all. If we look at the introduction of new technologies in any workplace, a number of key themes will emerge. First amongst these themes is the nature of technological innovation. In many cases, the rail industry has not yet experienced the relentless technological innovation that has characterised other industries (such as aviation) over the last 50 years. If a freight locomotive driver[1] were to go into a cab today compared with 50 years ago, the differences would probably be significantly less than for an aviator going into a cockpit. However, the rate of change in the rail industry in the

---

1   Different countries use different terms to describe the person who controls a train and the person who coordinates train movements in the network. Throughout this book we use the term train driver in a general way to refer to any person who drives a train (which includes engineers, train operators and so on) and the term train controller is used to refer to anyone who controls the movement of trains in the network (which includes signallers, dispatchers, area controllers, service controllers and so on).

next few decades is likely to be significantly greater compared to that of previous years. As such, the issues that may arise from rapid technological change and the potential for unintended or perverse outcomes might well increase exponentially.

It seems reasonable to predict that changes in the rail industry will be similar to those experienced by industries that have already undergone significant innovation, such as aviation. This provides some opportunity to learn from others' experience. If we look at aviation over the last few decades, the major technological changes (from a human factors perspective) have focused on improving safety and efficiency through automation and, where this is not possible, augmented cognition. It seems likely that similar changes will also occur in the rail industry in the near future. The case for increasing automation in rail operations is based on the idea that a very significant proportion of accidents and injuries result from human error (e.g. Reason 1990). Similarly, the efficiency demands in an operating environment where fuel costs are critical mean that even small improvements in driving performance can carry significant cost consequences at the organisational level.

For organisations dominated by engineering cultures, de-looping the human from these systems or providing automated decision support carries an intuitive appeal. It is argued that if humans are the source of the error, eliminating them or reducing our reliance on their judgment will make the system more efficient and/or safer. This leads to the design of new technologies that seek to eliminate the 'weaknesses' of humans in the system so that the system will, to paraphrase the NASA imperative, operate faster, safer, cheaper. However, the literature on introducing new technology suggests that while some benefits will undoubtedly occur, it is generally the case that new and different problems emerge to replace the old problems (McLeod, Walker and Mills, 2005; Sarter, Woods and Billings, 1997). For example, in highly automated systems which are designed to reduce human error from slips and lapses, it can be more challenging for the operator to maintain a good understanding of how the technology is functioning, which can lead to inappropriate control inputs and operational errors (Sarter et al., 1997). Another issue that has been highlighted with automated systems is the potential for the operator to place too much trust in the technology and disengage with the task (Parasuraman, Molloy and Singh, 1993). As we move to an increasingly automated rail industry, it is important to fully consider how the introduction of new technology changes rather than reduces errors, and how systems need to adapt in order to ensure that errors do not translate into accidents. This is an important theme of the book and one that each chapter is designed to address.

One of the focal themes of this book is the question of why the unintended occurs; what do we know about it already in other domains and how might one minimise the likelihood of it in the rail environment? In simple terms, the authors of the chapters in this book see the unintended as the consequences of overly narrow disciplinary focus on a problem. For example, the introduction of computer screens into the cab may provide additional information to support decision-making around the better use of momentum in achieving fuel efficiencies. However, it also means that operators are 'heads down' more often and less able to detect unpredictable changes in the external environment that are still dependent on being 'heads up', after all, railways are very much an open system. In many cases, the providers of highly technical solutions may lack experience or familiarity with the total task or the ways in which the task is conceptualised and executed by the operator as distinct from an engineer. In some cases it may reflect the different value systems that underlie the different disciplinary perspectives.

One of the key forces shaping the introduction of new technologies in rail is the sheer size or complexity of the rail network in many countries, particularly in freight operations. The need for greater energy efficiency, shorter separations between trains and greater volumes will mean that the complexity of optimising tasks safely may well exceed the capacity of a human decision-making system – especially at 5 o'clock in the morning! Increasingly we are providing workers with more cognitively dense decision spaces. As technology proliferates the operators will need to quickly process multiple sources of complex information in order to make decisions. In many cases we provide information on the basis of 'more is better' or because we do not want to leave it out in case it might have been useful and we are held accountable for not providing available information. However, there is a point at which additional information may be counterproductive (Roy, Breton and Rousseau, 2007).

The first problem is one of perception. As the amount of information increases, people struggle to see the critical dimensions of task performance due to low signal to noise ratios associated with high levels of distracting information. That is, important information relevant to a task can get lost in the mass of information that is presented to a user. The second problem is one of comprehension as the large amount of information makes it difficult for people to build and maintain an adequate mental model of the ongoing situation. Important questions that need to be asked about the provision of information to human operators are: What is the optimal information density for the decision space associated with specific tasks in the rail industry? How best can we display this on a monitor? At what point does additional information carry significant risk of distracting from, rather than enhancing, task performance? How do we determine this? How does the information provided by the technology support the development and maintenance of the operator's mental model? The answers to such questions shape the way we respond to the challenge of developing technologies that allow the operator to develop a good understanding of their operational world, given the increasing complexity of information provided by multiple interacting systems in the context of possibly degraded human functioning. Chapters 2, 7 and 10 address some of the theoretical issues that need to be considered in order to answer such questions along with some clear, simple techniques for determining this information at the task level in a specific organisation.

As the trend continues towards maximising network capacities, and the space separating trains is becoming less, drivers and train controllers must develop more sophisticated and time-sensitive mental models in order to execute the driving task as safely and effectively as possible. As a consequence, traditional 'foot plate' training programs may well not train staff sufficiently quickly or enable skill or route knowledge currency to be maintained.

To manage this we have seen a significant shift toward simulator-based training in recent years. According to the advocates of these systems, simulated environments can be used to accelerate training and to ensure currency where real-world currency is difficult to maintain. Experience tells us that these systems have significant benefits but are not necessarily interchangeable with a real-world training environment. What are the effects of significantly altering the relative frequency of situations in a simulated operational environment? Does frequent exposure to danger in a simulator lead to complacency and/or a false sense of security? Does route knowledge developed in a simulated environment translate to the real world? How does the lack of consequence in simulated environments influence risk-taking in the real world? Chapters 8 and 9 deal with these issues in detail.

While a technology may be conceived in isolation, the introduction of a technology into a train cab or railway control room occurs in a broader social, psychological and industrial/political context. The consideration of such factors, which may be outside the scope or experience of a technology solution provider, may determine the ultimate success of a new technology. There are many cases where new technologies can de-skill a task from those whose identity and seniority is defined by their ability to execute it themselves.

For example, when the FuelMiser technology for optimising driving strategies was introduced in the late 1990s in Australia, some train drivers felt devalued by the idea that years of accumulated expertise could readily be replaced by a computer algorithm developed by a physicist who had never driven a train. The algorithm, the screen technology and the way it was initially operationalised in the cab rendered the drivers redundant, and in their terms, they were now nothing more than 'horizontal elevator operators'. Moreover, the potential to automate the entire operational process using a remote operations approach was not lost on them. Not surprisingly, the inappropriate social and industrial contextualisation of the technology subverted its introduction and the vendors struggled to gain drivers' and unions' acceptance of the technology.

Perversely, when the same information was provided as an electronic equivalent of the old 'coasting boards' and it was suggested that the new technology merely provided a competitive benchmark against which the drivers could compete to demonstrate who was better in the field – amongst the drivers or against the computer – acceptance was much better. As the drivers built trust in the new technology, the technology was increasingly relied upon when drivers were in difficult, undulating territory or aware of the potential effects of fatigue on route planning and fuel economy. This shows the critical importance of non-technical aspects of new technology, and how a very clear understanding of how best to shape the way in which a technology will be socially constructed is critical to gaining support for and ensuring that the potential benefits of new technologies are realised. Chapter 3 discusses the issue of acceptance and resistance to technology.

Perhaps one of the biggest challenges facing the rail industry in this area is the question of participative design. Given the capital costs associated with new technologies in the rail industry and the global nature of the business, very few technologies can be designed from the bottom up – especially in smaller countries where the purchasing volumes are unlikely to enable significant redesign. In these cases, new purchasers will typically only have the capacity to influence minor aspects of the technology. In small markets, like Australia and Canada, purchases are likely to be based on modifying pre-existing technologies to local needs or conditions. Larger suppliers, whose business model is predicated on supplying the US, Indian or Chinese markets, are likely to have already constrained many of the design options available.

In this context, the dialogue around the introduction of a new technology may be more focused on identifying potential strengths and weaknesses and adapting the system into which pre-existing technologies are to be placed. This will require the capacity to anticipate the ways in which the new technology will influence the 'organisational ecosystem' into which it has been introduced and to ensure that the opportunities for the inadvertent consequences are minimised. Again, the chapters on consultative methodologies (chapters 4, 5, 6 and 7), both within the end-user and subject-matter expert communities, have addressed this issue and provide clear practical advice on how to ensure that an organisation minimises the likelihood of unanticipated outcomes.

However much care is exercised in anticipating the likely impacts of a new technology, it is unlikely that they can all be delineated and mitigated *ad hoc*. In many cases, post implementation surveillance will be a critical element of ensuring that issues are identified and mitigated quickly, albeit *post hoc*. Despite the application of risk assessment protocols, it is very difficult to anticipate all of the consequences of a new technology and sometimes we are forced to 'suck it and see': That is, the Rumsfeldian 'unknown unknowns' or the lack of knowledge about what we do not know can often loom large. Moreover, political and economic imperatives can often render careful and thoughtful *ad hoc* deliberation a theoretical nicety to which we can aspire but that we rarely attain. In these cases, the pressure to implement needs to be offset by a stronger emphasis on outcome evaluation. The evaluation of a new technology may occur in what has been typically referred to as an 'action research' or 'continuous improvement' paradigm.

One of the case studies presented in chapter 9 demonstrates the power of a highly focused human factors analysis of a new technology around fatigue risk management. A rail operator had introduced a software package for quantifying the level of sleep opportunity (a proxy for work-related fatigue) associated with a roster/schedule and wanted to understand whether the measure was associated with changes in driving safety and efficiency. This very detailed analysis indicates the usefulness of post-implementation evaluation and the way in which a strong focus on *post hoc* assessment can ensure that a new technology has, or has not, delivered on its promised benefits.

In closing, it is important to understand what this book is trying to achieve. It does not claim that all new technologies can be designed, implemented and evaluated systematically so as to completely eliminate error or inadvertent consequence. It does not claim that there are rigorous, well-standardised theories and methodologies for assessing new technologies. What it does claim to do is to provide a useful overview of how one might think through the issues associated with introducing a new technology so that the likelihood of error and unintended consequences is reduced. In providing this overview we have attempted to illustrate the complexity of the issues around successfully introducing a new technology and the need to approach the issue with a collaborative, multi-disciplinary perspective. In particular the approaches taken in the book show the limitations of a narrow technical perspective on new technologies.

Most importantly the book illustrates the centrality of non-technical factors and the importance of participative design principles being applied both before and after the implementation of a new technology. At a practical level it shows the benefits of being able to make fast mistakes and provides some useful ways to ensure most of the mistakes can be identified and the operation made much faster and cheaper than has been the case when the issue is not addressed. And lastly, it shows that a little forethought is a powerful thing, that is, a stitch in time saves nine!

References

McLeod, R.W., Walker, G.H., and Mills, A. (2005). Assessing the human factors risks in extending the use of AWS. In J.R. Wilson, B. Norris, T. Clarke and A. Mills (eds.), *Rail Human Factors: Supporting the Integrated Railway* (pp. 109–19). Aldershot: Ashgate Publishing.

Parasuraman, R., Molloy, R., and Singh, I.L. (1993). Performance consequences of automation-induced complacency. *International Journal of Aviation Psychology, 3*, 1–23.

Reason, J. (1990). *Human Error*. New York: Cambridge University Press.

Roy, J., Breton, R., and Rousseau, R. (2007). Situation awareness and analysis models. In E. Bossé, J. Roy and S. Wark (eds.), *Concepts, Models, and Tools for Information Fusion* (pp. 27–67). Boston: Artech House.

Sarter, N.B., Woods, D.D., and Billings, C.E. (1997). Automation surprises. In G. Salvendy (ed.), *Handbook of Human Factors and Ergonomics (2nd edn)*. New York: John Wiley & Sons, Inc.

Chapter 2

Key Technology-Related
Human Factors Issues

Chris Bearman

Central Queensland University, Appleton Institute, Adelaide, Australia

Introduction

Introducing new technology generally has the potential to improve safety and/ or efficiency. For example, Positive Train Control is designed to both enhance operations and increase safety. Positive Train Control is designed to protect the train from driver error by applying the brakes if the system thinks the driver will over-speed, pass a signal showing a red aspect or operate outside limits set by train controllers (Wreathall, Woods, Bing and Christoffersen, 2007). This seeks to reduce driver errors that lead to worker injuries, harm to the general public and significant economic losses for the rail operators (Wreathall, Woods et al., 2007). According to Wreathall, Woods et al. (2007), the US National Transportation Safety Board has identified over 100 collisions that could have been avoided by a full-function Positive Train Control system. The installation of a similar system in the UK (known as the Train Protection Warning System, TPWS) that intervenes if the train is predicted to over-speed or fail to stop at a red signal has led to a 22 per cent reduction in signals passed at danger (SPADs) and an estimated 86 per cent reduction in overall risk (Rail Accident Investigation Branch, 2008). This presents an impressive case for the installation of such systems in the rail industry.

However, introducing new technology is not without risk. There have been several recent rail incidents relating to human interaction with technology. For example, at Didcot Junction (UK) a driver automatically cancelled two auditory warnings (from an Automatic Warning System) that the next signal was red, without checking on the aspect of the signal, leading to a red signal being passed and near-collision with another train (Rail Accident Investigation Branch, 2008). In a similar incident at Purley (UK) a driver incorrectly assumed that a TPWS warning that he received was erroneous and overrode it, continuing up the line against a red signal (Rail Accident Investigation Branch, 2007b). Technology that is not central to the control of the train has also been implicated in serious rail incidents. For example, near Chatsworth in California (USA) two trains collided, leading to 25 fatalities and 135 injuries because one of the drivers missed a signal showing a red aspect, most likely because he was using a mobile phone (National Transportation Safety Board, 2010).

In situations such as those outlined above, it is usually assumed that the driver is at fault for neglecting to properly assess the situation or for engaging in inappropriate actions. While this may be true in some cases, in many others it may represent only part of a much more complex story. The reality of operating in a complex environment (such as a train cab or train control room) is that there are often awkward to use interfaces, frequent false alarms, ambiguous information, and multiple warnings that do not require action. Under these conditions drivers can sometimes build an understanding of a situation that is incorrect, leading to inappropriate actions. It is important then to understand the potential technology-related human factors issues associated with introducing new technology so that the potential for error can be determined and a thorough evaluation of the risks can be made.

A wide range of technology-related human factors issues have been discussed in relation to rail technologies (c.f. Rose and Bearman, 2012; Wreathall, Roth, Bley and Multer, 2007; Young, Stanton and Walker, 2006). While the human factors issues that have been identified so far are specific to particular technologies, a more general set of issues can be derived that together provide an introduction to some of the basic technology-related human factors issues (see Table 2.1). These human factors issues consider both the way that people interact with technology and the impact that a technology has on people's ongoing work activities.

Table 2.1   Key technology-related human factors issues

Inadequate operator understanding of the technology
Sub-optimal physical design or location of the technology
Sub-optimal information provision or feedback
Distraction
Attenuation to alarms
Failing to act on an alarm
Problems transitioning between different modes

Together these issues form the starting point for evaluating a technology, or alternately a list of basic issues to discuss with the salesman. The following sections provide a more in-depth discussion of these human factors issues together with examples from situations where these issues have led to accidents.

Understanding the Technology

In order to be able to use a technology effectively in a variety of different situations the operator must have a good understanding of how that technology works. When an operator learns about a system they build a mental model about how that technology functions, the different modes in which it can operate and conditions under which certain system behaviours would be expected to occur. This learning is typically developed through short training courses, using the system in the course of their job and through explaining the system to others. However, such learning is rarely comprehensive and it is likely that the operator's mental model of how the technology functions will contain idiosyncratic gaps and/or incorrect information, even after extensive experience (McClumpha and James, 1994; Sarter and Woods, 1994). The presence of these gaps and incorrect information can lead to errors inputting information into the system and incorrect use of the technology, particularly when it is used outside the normal range of operations.

Gaps and incorrect information may be present in an operator's mental model of a technology even if that technology provides a critical safety feature. In the collision between two trains at Illabo (Australia) (Australian Transport Safety Bureau, 2008), the train crew incorrectly believed that they had activated an important communication system. The train crew's failure to activate the communication system prevented train controllers from being able to warn the train crew of the impending collision (see Case Study 2.1 for more information about this accident).

Case Study 2.1 – Operators may not understand how a system works, even if it is a critical safety system! (taken from Australian Transport Safety Bureau, 2008)

In Australia, near Illabo in New South Wales, a freight train collided with an overturned semi-trailer that was blocking a level crossing. The train crew were treated for shock but fortunately there were no serious injuries. A train controller had attempted to contact the train crew to warn them of the level crossing obstruction on four occasions but had not been able to reach them. The train controller was unsuccessful in his attempts to communicate with the train crew because the radio he was attempting to contact them on was in fact in an unmanned trailing locomotive. The train crew had not correctly activated the radio (which was supposed to be the primary communication device) in the leading locomotive which they were occupying. Both train drivers stated that they believed that the radio had been switched on and registered on the network because the radio had chirped and an orange radio indicator light had flashed when they were travelling through a nearby town. Both the train crew and two separate train controllers had the opportunity to correct the error but did not do so.

A section from the report into the derailment at Cajon Junction (USA) (National Transportation Safety Board, 1996) highlights some of the issues around inadequate understanding of an important safety system. In this report the train driver reported that he had difficulty setting a Two-Way End of Train Device (ETD) which facilitates braking on steep inclines. The driver claimed to know the proper process of arming the ETD and had read and understood the procedures associated with arming the system. However, the driver had acquired his knowledge of the arming procedure informally 'on the job' and not in association with a comprehensive competency-based training program. The driver said that he experienced only a 50 per cent success rate in arming ETDs and the ETD system did not provide good feedback on why it could not be armed. While in the case of the derailment at Cajon Junction the lack of understanding of the technology did not directly contribute to the accident, this example does demonstrate that drivers may have incomplete knowledge of how an important system works and why it is providing the response that it does.

Operator training in a new technology should aim to be comprehensive, allow exploration of the system by the user and encompass sessions where the focus is on rectifying potential gaps in the drivers' mental models of the system. Training and skills transfer are discussed further in chapter 8. One useful technique for identifying and rectifying gaps in mental models is the process of self-explanation (Chi, 2000). Self-explanation (or explaining information to oneself) has been found to lead to improved recall and use of information compared to other cognitive processes (such as summarisation) (Bearman, Ormerod, Ball and Deptula, 2011; Chi, DeLeeuw, Chiu and LaVancher, 1994; Neuman and Schwarz, 1998). This effect appears to be due to self-explanation leading to the identification of gaps in knowledge and the formation of inferences to cover the gaps, which leads to the revision or reorganisation of the contents of the mental model (Chi, 2000). The process of self-explanation may therefore be a useful way to facilitate drivers' development of a more comprehensive understanding of a new technology.

Physical Design, Location Information Provision and Feedback

The physical design of a technology (such as the ease of using the interface and the location of aspects of that interface in relation to the operator) plays a large part in how easy it is to use/understand and the errors that can be made in using it. Even if a technology is well designed, the integration with other equipment in the train cab or control room may be a problem. Train cabs and train control rooms have limited space and new technology may need to be placed in non-optimal positions.

One example of the poor placement of a technology's interface leading to the removal of a safety barrier can be found in the derailment at Carlstadt (Canada) (Transportation Safety Board of Canada, 2003). The need to reach above and to the left to activate a radio contributed to the driver inadvertently selecting the wrong communication channel. This led to the drivers missing a hot-box detector alarm that would have warned the drivers of the train fault, which resulted in the train derailing.

On the basis of an extensive review of human factors research, Multer, Rudich and Yearwood (1998) have produced a comprehensive set of guidelines for evaluating the physical design of the cab environment, cab layout and workstation design. In terms of workstation design the guidelines consider the way that people interact with control input devices and perceive information from displays, as well as considering the most appropriate locations for displays and devices given the limitations of human physiology. The interested reader is directed to this publication for more information.

An important part of the design of an interface is the way in which it presents information and feedback to the operator. The technology needs to present sufficient information to the operator to allow the operator to understand the situation but not so much that they become overloaded or the important information is lost in unnecessary clutter. When a technology replaces one or more of the operators' work roles it is especially important for the technology to provide information about what it is currently doing and what it will do in the immediate future. When the feedback from the system is inadequate the operator will have a degraded understanding of what the system is doing. This means that the operator will be unlikely to identify any inconsistencies between how they think the system is operating and how it actually is operating and will be unable to predict what the system will do next. This leads to the operator being out of the loop with respect to the functioning of the technology and to automation surprises, where the system behaves in unexpected or unpredictable ways (Sarter, Woods and Billings, 1997).

Poor feedback from technology has contributed to a number of rail incidents. In an incident at Desborough in the UK (Rail Accident Investigation Branch, 2007a), an automatic brake application was cancelled three times by the driver because the system presented only ambiguous information in the driver's primary field of vision. While a clearer warning light was provided, this warning was on a panel that the driver only looked at when the train was stopped at a platform. This led the driver to make an erroneous judgement to proceed instead of stopping immediately as was warranted by the situation (an open passenger door). More information about this incident is provided in Case Study 2.2.

In a train control environment, controllers in Adelaide (Australia) did not initially receive an unambiguous warning that a train had passed a signal at danger and was proceeding up the wrong track (Australian Transport Safety Bureau, 2007), and following a SPAD at Fisherman Islands (Australia) the train controller who was the first point of contact for emergencies did not have any indication of where the train was or what aspect one of the signals was displaying (Australian Transport Safety Bureau, 2005).

Humans are good at adapting to technology and are able to effectively use systems that are awkward to use (either because they provide an unclear interface or ambiguous feedback) (Endsley, Bolté and Jones, 2003). However, while the human can overcome the limitations of poor technology they will not necessarily develop an adequate understanding of how that system works, which may lead to errors when that system needs to be used outside of the normal range of activities (such as under time constraints or emergency situations). In the derailment at Chiltern (Australia) 'despite both drivers being trained in the use of emergency communications equipment, under the pressure of an emergency situation neither driver was sure how the equipment worked' (Australian Transport Safety Bureau,

---

Case Study 2.2 – Emergency situations are sometimes not clearly indicated by a technology (taken from Rail Accident Investigation Branch, 2007a)

In the UK, a passenger train on a mainline service from London to Sheffield continued along the track for 5 miles with one of the passenger doors open. This presented a 'real and unprotected risk to those on board the train' (p. 7). When the door came open the train automatically applied a brake. The only indication from the train about the cause of this brake that the driver could see was the illumination of the 'pass comm/door activated light'. This light can illuminate for a number of reasons, such as someone activating a door, a door malfunction, or one of the passengers using an emergency brake handle. There was also no information available to the driver on the train management system since the open door was not logged as an anomaly on this system. However, the train did provide another indication about the cause of the problem on a panel that was not in the driver's primary visual field. This light is not in the driver's primary field of vision since its primary function is to indicate the state of the doors when the train is stationary. When the door came open the 'door close/locked' light would have extinguished. If the driver had seen this, it would have indicated to him that the brake application was caused by an open door. Since there was no fault logged in the train management system and no communication had been initiated with a passenger, the driver most likely assumed that the warning system was faulty and he overrode the automatic brake application with the intention of stopping at the first suitable location, instead of stopping the train immediately as was warranted by having an open passenger door on the train.

2004, p. 22). This lack of understanding of the emergency equipment may have contributed to the unsuccessful attempts of the drivers of the derailed train to warn the driver of a second train approaching the site about the derailment. It is important then to produce simple, easy to use technology that is intuitive to the user under the pressures of an emergency.

The physical design of the system and the information and feedback that it provides to users may contribute to a number of other technology-related human factors issues, such as how easy it is for the operator to develop a good mental understanding of the technology, how distracting that technology is and the extent to which warnings are clear and unambiguous. Hence, it is particularly important to evaluate the physical design of a technology at an early stage of the design process and to consider how the technology will interact with other existing technologies. Early evaluation of a technology ensures that aspects of the system that are prone to human factors issues will not be carried forward into later stages of the process where they are harder and more expensive to rectify. A range of options exists for evaluating a technology throughout the design process (such as mock-ups, simulation, observation, task analysis, Future Inquiry workshops and interviews). These methods are discussed in chapters 4 to 10.

Distraction

An important consideration when introducing a new technology into the train cab or train control room is whether that technology will be distracting. Distraction is a particular problem when the technology is first introduced and people want to play with it to see what it does. When a driver is distracted the focus of their attention is on the technology, the information it is providing or on conversations about (or using) the technology rather than on the primary task of driving the train. This can lead to the driver failing to perceive important information about the operation, failing to comprehend information if it is perceived or failing to predict what will occur in the future based on this information (Endsley et al., 2003). While it is unreasonable to expect that a person will have their full attention on driving the train all the time, the presence of a competing focus for attention means that the driver will have a harder time focusing on the primary task of driving the train.

As expected, distraction and reduced awareness of the current situation can lead to serious consequences. In the accident at Silver Spring (USA) (National Transportation Safety Board, 1997), the train crew needed to pay attention to a wayside information system which broadcasted information over a common communication channel. When two trains passed over a monitor in close succession, the crews had to pay close attention to the system to determine which information related to their train. This, added to the already high workload of the crew, degraded their situation awareness of the route, causing the crew to forget that they had passed a caution signal, which led to a signal passed at danger and a collision with another train.

Distraction can also result from the use of technology that is not central to the primary task of operating the equipment (e.g. driving the train). In particular, the use of communication devices has led to accidents. The use of mobile phones by train drivers was found to be a causal factor in the rail accidents at Clarendon (USA) (National Transportation Safety Board, 2003) and Chatsworth (USA) (National Transportation Safety Board, 2010). In the accident near Chatsworth in Southern California, USA, a passenger train collided head-on with a freight train after the passenger train had passed a red signal. The investigation report into the accident concluded that the probable cause of the collision was the driver of the passenger train failing to 'observe and appropriately respond to the red signal aspect ... because he was engaged in prohibited use of a wireless device, specifically text messaging, that distracted him from his duties' (National Transportation Safety Board, 2010, pp. vii).

At Clarendon in Texas, USA, a coal train collided head-on with a freight train after the coal train failed to stop at a point designated by the train controller. At the time that the train controller's instructions should have been attended to, the train driver was holding a conversation on his personal mobile phone. This conversation likely diverted enough of the train driver's attention away from the train controller's instructions so that he missed the fact that he needed to wait for the other freight train to pass before proceeding along the track (National Transportation Safety Board, 2003).

While use of communication devices by the drivers in both of these accidents was for personal rather than work purposes, it should be noted that several technologies have been developed for train communications that closely resemble mobile phones. It is therefore important to consider the issues around possible distraction created by the use of communication devices by drivers.

Research in the automotive industry suggests that using mobile phones represents a significant hazard. In particular, studies have found that using a mobile phone while driving can cause impairment in the perception and response to relevant information, even when the information relates to potentially dangerous situations (Caird, Willness, Steel and Scialfa, 2008; Kass, Cole and Stanney, 2007). A meta-analysis on driving performance and mobile phone use conducted by Caird et al. (2008) found a similar level of impairment when drivers used hands-free devices compared to hand-held phones, suggesting that the impairment in performance is based on the cognitive demands of conducting a conversation rather than a specific device (Maciej, Nitsch and Vollrath, 2011). While this research was done in the automotive industry rather than the rail industry, the findings suggest that it is vitally important to thoroughly evaluate the impact of communication systems before they are implemented.

One way of preventing operator distraction is to have periods of time when non-essential communication is restricted so that the operator focuses on critical aspects of the task. However, numerous train observations conducted by the author and colleagues have found that drivers frequently ignore these communication-restricted periods of time. Communication-restricted periods of time are therefore a weak solution to the problem, and other methods of ensuring that distraction does not occur should be considered (such as blacking out the technology during critical zones). There are a number of methods for assessing the potential for distraction that a technology possesses. These methods will be discussed further in the chapters on qualitative methods (chapter 5), task analysis (chapter 7) and situation awareness (chapter 10).

Response to Alarms

It is often considered to be inconceivable that an alarm will fail to trigger an appropriate response by a driver or train controller (Hall, 1999). However, alarms occur in an ongoing stream of events in the operational environment, where the operator is constantly building an understanding of their current situation and responding to external stimuli. In this context, there are two key reasons why an operator may not respond appropriately to an alarm. The first is that response to frequent alarms can become automatic and the second is that warnings are incorporated into the operator's current understanding of the situation before they take action (Brenitz, 1983; Endsley et al., 2003; McLeod, Walker and Mills, 2005).

It has long been known that vigilance systems that require a response to an alarm develop patterns of automatic (stimulus-response) actions in train drivers where they automatically respond to the warning regardless of their state of arousal (Dorrian, Lamond, Kozuchowski and Dawson, 2008). One story related to the author by a train driver about vigilance systems is that when the train driver's alarm goes off in the morning to wake him up to go to work, he reaches out to hit the vigilance alarm button.

Systems that provide multiple warnings may also lead to driver attenuation to warnings, where the warnings are either assumed by the driver to be meaningless or are cancelled 'automatically' before they are fully investigated (McLeod et al., 2005). This is a particular problem when the driver is expecting an alarm to sound. McLeod et al. (2005) investigated an Automated Warning System (AWS) which provided warnings of signal aspect to train drivers. McLeod et al. (2005) found that 56 per cent of drivers reported 'automatically' cancelling an AWS warning without mentally processing it, on more than one occasion. Further, on the basis of reaction time data, McLeod et al. (2005) found that the driver's response to the warning is highly learnt and/or anticipatory in nature.

Cancelling an alarm without performing appropriate actions has been cited in a number of accident reports and has occurred in both train driving and train control. In a SPAD and near collision at Didcot (UK) the driver cancelled two Automatic Warning System warnings without checking on the aspect of the signal ahead (Rail Accident Investigation Branch, 2008). In the SPAD at Fisherman Islands (Australia) a controller cancelled a SPAD alarm within 10 seconds of it sounding but made no attempt to contact either the train driver or the controllers who coordinate emergencies (Australian Transport Safety Bureau, 2005).

One of the things that an operator needs to consider when they hear an alarm is whether the alarm is real or spurious. When a driver experiences numerous false alarms from a technology, a real alarm is more likely to be interpreted as spurious, particularly if the event that the alarm is indicating is rare. Endsley et al. (2003) make the point that operators do not just blindly respond to an alarm. An alarm must be integrated with other information in the environment in order to understand the meaning of that alarm and how to react to it (Brenitz, 1983; Endsley et al., 2003). Even in systems that provide instructions on how to act and require immediate compliance (such as the Ground Proximity Warning System in aviation) the operator will consider the false alarm rate of the system and their current situation before responding (DeCelles, 1991).

When information from the environment can be construed by the operator as consistent with an interpretation of an alarm as spurious then it is more likely that the alarm will not be heeded. An example of this occurred in the SPAD at Purley Station in the UK (Rail Accident Investigation Branch, 2007b): When leaving the station a train driver received a TPWS intervention which applied the brakes. The driver overrode this intervention thinking that it was erroneous, like the previous two warnings he had received at that station. The driver's assessment of the situation was confirmed to him by his observation of a sequence of green lights up the track. Unfortunately the driver had positioned his train so that he was unable to see the signal that related to his platform and the green lights related to a train on the adjacent platform. The TPWS intervention had occurred as a result of the driver passing a red signal at the end of the platform.

It is generally assumed that implementing a technology that provides a warning to a driver will be sufficient to prevent incidents caused by driver error. However, there is a growing literature that suggests that this may not necessarily be the case. These findings indicate that the installation of a technology that has an alarm needs to be done carefully, with full consideration of how that alarm will be perceived by the driver. The methods discussed in chapters 4, 5, 7, 9 and 10 outline some ways to do this.

Transitioning Between Different Modes of the Technology

In large countries, such as Australia, Canada, China, India and the United States, it is likely that there will be areas where a technology operates and areas where it will not. This is particularly likely when the technology is being rolled out across a network (Young et al., 2006). Transitioning between the places where the technology is operational and places where the technology is not operational may lead to sudden increases in operator workload, confusion about the mode the technology is operating in and an increase in the risk of error through inappropriate reliance on the technology (Wreathall, Roth et al., 2007). Transitions into and out of areas covered by the new technology may also require drivers to switch between driving techniques so that sometimes they are driving heads-up (route focused) and sometimes they are driving heads-down (instrument focused) (Porter, 2002). The effects of such transitions are currently unknown and need to be subject to further research.

Concluding Thoughts

The development of new technology is an important source of innovation for rail organisations. New technologies promise considerable benefits in terms of enhanced capabilities, more efficient operations and improved safety. However, the introduction of technologies can also create unforeseen problems leading to disruptions to rail operations and compromised safety. This chapter has introduced a number of technology-related human factors issues that need to be considered when introducing a technology. Together these issues form the starting point for evaluating a technology, or alternately a list of basic issues to discuss with the salesman.

It should be noted that a well-designed technology without any obvious flaws might still interact in unexpected ways with other technologies in the train cab, particularly if the technologies provide an alarm to the driver. This means that an evaluation of the technology in the context in which it will be operating is essential. Ideally the evaluation of a new technology should begin in conjunction with the tender process and early design phases, so that costly revisions to the technology down the line are avoided. Chapter 4 discusses this in more detail. Chapters 4 to 10 present a range of methods that can provide information about each of the potential issues.

Even if a technology does not possess any serious human factors issues, it is still possible that it will be rejected by the people who will be using that technology. The reasons why people resist new technology, some of the forms this resistance takes and ways to reduce negative resistance are discussed in the next chapter.

Acknowledgements

The author is grateful to the CRC for Rail Innovation (established and supported under the Australian Government's Cooperative Research Centres program) for the funding of this research (R2.102 – Human Factors Analytic Tools Project). I would also like to thank Drew Dawson and Jillian Dorrian for their comments on an earlier version of this manuscript and Levi McCusker for his work in identifying the rail accidents. This chapter was written in part on a retreat held in the Anangu Pitjantjatjara Yankunytjatjara (APY) lands organised by Drew Dawson.

References

Australian Transport Safety Bureau. (2004). *Derailment of Pacific National Freight Train 1SP2N and the Subsequent Collision of V/Line Passenger Train 8318 near Chiltern, Victoria. 16 March 2003* (Rail Safety Investigation 2003/02). Canberra: Australian Transport Safety Bureau.

Australian Transport Safety Bureau. (2005). *Signal FS66 Passed at Danger, Freight Train 8868. Fisherman Islands, Queensland. 20 September 2004* (Rail Occurrence Investigation 2004/004). Canberra: Australian Transport Safety Bureau.

Australian Transport Safety Bureau. (2007). *Signal 161 Passed at Danger TransAdelaide Passenger Train H307, Adelaide, South Australia. 28 March 2006* (Rail Occurrence Investigation 2006/003). Canberra: Australian Transport Safety Bureau.

Australian Transport Safety Bureau. (2008). *Collision Between Freight Train 9351 and an Overturned Semi-Trailer. Illabo, New South Wales. 2 November 2006* (Rail Occurrence Investigation 2006/013). Canberra: Australian Transport Safety Bureau.

Bearman, C.R., Ormerod, T.C., Ball. L.J., and Deptula, D. (2011). Explaining away the negative effects of evaluation on analogical transfer. *The Quarterly Journal of Experimental Psychology, 65*(5), 942–59.

Brenitz, S. (1983). *Cry Wolf: The Psychology of False Alarms*. Hillsdale, NJ: Lawrence Erlbaum.

Caird, J.K., Willness, C.R., Steel, P., and Scialfa, C. (2008). A meta-analysis of the effects of cell phones on drivers' performance. *Accident Analysis and Prevention, 40*, 1282–93.

Chi, M.T.H. (2000). Self-explaining: the dual processes of generating inferences and repairing mental models. In R. Glaser (ed.), *Advances in Instructional Psychology* (pp. 161–238). Mahwah, NJ: Lawrence Erlbaum Associates.

Chi, M.T.H., de Leeuw, N., Chiu, M., and LaVancher, C. (1994). Eliciting self-explanations improves understanding. *Cognitive Science, 18*, 439–77.

DeCelles, J.L. (1991). *The Delayed GPWS Response Syndrome*. Herndon, VA: Aviation Research and Education Foundation.

Dorrian, J., Lamond, N., Kozuchowski, K., and Dawson, D. (2008). The Driver Vigilance Telemetric Control System (DVTCS): investigating sensitivity to experimentally-induced sleep loss and fatigue. *Behavioural Research Methods, 40*, 1016–25.

Endsley, M.R., Bolté, B., and Jones, D.G. (2003). *Designing for Situation Awareness: An Approach to User-Centred Design*. London: Taylor and Francis.

Hall, S. (1999). *Hidden Dangers: Railway Safety in the Era of Privatization*. Shepperton: Ian Allen.

Kass, S.J., Cole, K.S., and Stanney, C.J. (2007). Effects of distraction and experience on situation awareness and simulated driving. *Transportation Research Part F: Traffic Psychology and Behaviour, 10*, 321–9.

Maciej, J., Nitsch, M., and Vollrath, M. (2011). Conversing while driving: the importance of visual information for conversation modulation. *Transportation Research Part F, 14*, 512–24.

McClumpha, A., and James, M. (1994). Understanding automated aircraft. In M. Mouloua and R. Parasuraman (eds.), *Human Performance in Automated Systems: Current Research and Trends* (pp. 183–90). Hillsdale, NJ: Lawrence Erlbaum Associates.

McLeod, R.W., Walker, G.H., and Mills, A. (2005). Assessing the human factors risks in extending the use of AWS. In J.R. Wilson, B. Norris, T. Clarke and A. Mills (eds.), *Rail Human Factors: Supporting the Integrated Railway* (pp. 109–19). Aldershot: Ashgate Publishing.

Multer, J., Rudich, R., and Yearwood, K. (1998). *Human Factors Guidelines for Locomotive Cabs* (DOT/FRA/ORD-98/03). Washington, DC: Federal Railroad Administration.

National Transportation Safety Board (1996). *Derailment of Freight Train H-Balti-31, Atchison, Topeka and Santa Fe Railway Company Near Cajon Junction, California on 1 February 1996* (Railroad Accident Report NTSB/RAR-96/05). Washington, DC: National Transportation Safety Board.

National Transportation Safety Board. (1997). *Collision and Derailment of Maryland Rail Commuter MARC Train 286 and National Railroad Passenger Corporation Amtrack Train 29 near Silver Spring, Maryland. 16 February 1996* (Railroad Accident Report NTSB/RAR-97/02). Washington, DC: National Transportation Safety Board.

National Transportation Safety Board. (2003). *Collision of Two Burlington Northern Santa Fe Freight Trains Near Clarendon, Texas. 28 May 2002* (Railroad Accident Report NTSB/RAR-03/01). Washington, DC: National Transportation Safety Board.

National Transportation Safety Board. (2010). *Collision of Metrolink Train 111 With Union Pacific Train LOF65-12, Chatsworth, California. 12 September 2008* (Railroad Accident Report NTSB/RAR-10/01). Washington, DC: National Transportation Safety Board.

Neuman, Y. and Schwarz, B. (1998). Is self-explanation while solving problems helpful? The case of analogical problem solving. *British Journal of Educational Psychology, 68*, 15–24.

Porter, D. (2002). Implementing ERTMS in the UK: human factors implications for the train driver. *International Railway Safety Conference, November 2002, Tokyo, Japan.* Retrieved on 6 June 2013 from http://www.intlrailsafety.com/Tokyo/3-2Tokyo_FullPaper.doc.

Rail Accident Investigation Branch. (2007a). *Passenger Door Open on a Moving Train Near Desborough. 10 June 2006.* London: Rail Accident Investigation Branch.

Rail Accident Investigation Branch. (2007b). *Signal T172 Passed at Danger at Purley station, Surrey. 18 August 2006.* London: Rail Accident Investigation Branch.

Rail Accident Investigation Branch. (2008). *Signal Passed at Danger and Subsequent Near Miss at Didcot North Junction 22 August 2007.* London: Rail Accident Investigation Branch.

Rose, J., and Bearman, C. (2012). Making effective use of task analysis to identify human factors issues in new rail technology. *Applied Ergonomics, 43*, 614–24.

Sarter, N.B., and Woods, D.D. (1994). Pilot interaction with cockpit automation II: an experimental study of pilots' models and awareness of the flight management and guidance system. *International Journal of Aviation Psychology, 41*, 1–28.

Sarter, N.B., Woods, D.D., and Billings, C.E. (1997) Automation surprises. In G. Salvendy (ed.), *Handbook of Human Factors and Ergonomics (2nd edn).* New York: John Wiley & Sons, Inc.

Transportation Safety Board of Canada. (2003). *Derailment Canadian Pacific Railway Freight Train No. 202-16 Mile 86.5, Kaministiquia Subdivision. Carlstadt, Ontario 19 October 2003* (Railway Investigation Report R03W0169). Gatineau, QC: Transportation Safety Board of Canada.

Wreathall, J., Roth, E., Bley, D., and Multer, J. (2007). *Human Factors Considerations in the Evaluation of Processor-Based Signal and Train Control Systems: Human Factors and Railroad Operations (DOT/FRA/ORD-07/07)*. Washington, DC: Federal Railroad Administration.

Wreathall, J., Woods, D.D., Bing, A.J., and Christoffersen, K. (2007). *Relative Risk of Workload Transitions in Positive Train Control*. Technical Reports (DOT/FRA/ORD-07/12), Washington, DC: Federal Railroad Administration.

Young, M., Stanton, N., and Walker, G. (2006). In loco intelligentia: human factors for the future European train driver. *International Journal of Industrial and Systems Engineering, 1*, 485–501.

Chapter 3
Resistance to Technology

Janette Rose
University of South Australia, Adelaide, Australia

Chris Bearman
Central Queensland University, Appleton Institute, Adelaide, Australia

Why Do People Resist New Technology

One of the main questions that people ask about resistance to technology is 'Why?' Why are people resistant to the implementation of new technology? There has been limited research in this area in the transportation domain, although researchers from other domains have looked at this question in slightly different ways, yielding a considerable body of literature on this issue. This research is placed into sections in this chapter that provide a partial answer to this question. These sections are: perceived usefulness and perceived ease of use of the technology; perceived impact on the user; individual characteristics; social influences; and organisational factors. While these sections provide a partial answer to the question of why people are resistant to technology, it should be pointed out that resistance to any given technology is likely to be a combination of these reasons for both individuals and groups. It is also important to keep in mind that resistance should not be discounted or ignored as some of the reasons for resistance are justified and could play an important role in ensuring safety and avoiding implementation of costly equipment that has little benefit to the organisation. The issue of beneficial resistance is considered in 'Is Resistance Always Unreasonable' below.

Perceived Usefulness and Perceived Ease of Use

One of the key determinants of whether people will accept a new technology or whether they will be resistant to its use is how useful the system is perceived to be and how easy it is to use. Davis (1986) has developed a highly influential and commonly used model of technology acceptance called the Technology Acceptance Model (TAM), which proposes that system design affects perceived ease of use and perceived usefulness. Perceived ease of use and perceived usefulness determine the user's attitude toward the system, which in turn predicts actual system use (and hence acceptance). Put simply, if a technology is to be accepted by operators or the end users, it must be perceived as being useful and

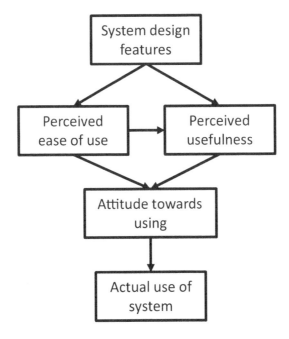

Figure 3.1   Technology Acceptance Model (adapted from Davis, 1986)

easy to use, and the specific design features will affect the users' perceptions of these attributes (see Figure 3.1 above).

A longitudinal study conducted by Davis, Bagozzi and Warshaw (1989) found evidence to support the assertion that technology use can be predicted from only three theoretical constructs, namely behavioural intention, perceived usefulness and perceived ease of use. In their study, university students were given a one-hour introduction to a word-processing program that would be available to use but was not compulsory. Immediately following the introductory session, students completed a questionnaire regarding their intentions, attitudes and beliefs about using the word-processing program (including questions relating to perceived usefulness and perceived ease of use). At the end of the 14-week semester, a second similar questionnaire was administered, with the inclusion of questions relating to actual usage. Results showed that after the introductory session, perceived usefulness and perceived ease of use were predictive of students' intention to use the program, however, at the end of the semester, only perceived usefulness was predictive of intention, with ease of use having only an indirect influence on intention via usefulness (Davis et al., 1989).

Further research has found that the relationships between perceived usefulness and perceived ease of use are complex and may vary between the contexts in which they are examined. Research by Davis et al. (1989) suggests that perceived usefulness is the key concept in predicting whether a technology is accepted, with ease of use seeming to become less important as the user becomes more familiar with the system. In contrast, other research by Robinson, Marshall and Stamps (2005) suggests that it is perceived ease of use, not perceived usefulness, that is directly related to intention to use a technology. This relationship led Robinson et al. to conclude that perceiving a technology as being easy to use leads to better performance, which in turn provides the user with a better opportunity to learn its usefulness. Robinson et al. (2005) also found that personal innovativeness and perceived level of availability of support services were positively related to perceived ease of use, suggesting an important role for support services in the acceptance of technology. Figure 3.2 shows the interaction of these different concepts with the original TAM model.

It is not clear why Davis et al. (1989), and Robinson et al. (2005) found different relationships between perceived usefulness and ease of use, although possible explanations may be related to differences between the technologies that were examined in the different studies and the domains in which they were introduced. It could be that the relationship between perceived usefulness and ease of use is dependent on the specific context in which the concepts are examined.

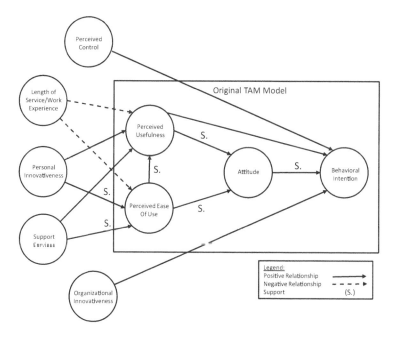

Figure 3.2   Structural model and hypotheses supported (Robinson et al., 2005; reprinted by permission of Elsevier)

While the exact relationship between perceived usefulness, ease of use and intention to use a new technology is not clear and may vary between different contexts, it seems that both usefulness and ease of use need to be carefully considered when developing a new technology. Further, the usefulness and ease of use of a technology need to be clearly demonstrated to users so that they can easily perceive the benefits. User involvement during the design process plays a key role in ensuring that a technology is clearly beneficial to the user and easy to use. If the users help to develop the technology and contribute to its design, it is more likely to be perceived by other users as being useful and easy to use since the technology is designed around the human rather than the human having to adapt to the technology (Sarter, Woods and Billings, 1997). Having said that, simply involving users does not guarantee acceptance. This is discussed further in 'Minimising Resistance' below.

Perceived Personal Impact

Another key influence on whether users will accept a new technology is the perceived impact that it has on their work role. In a summary of the major perspectives on technology and work organisation, Liker, Haddad and Karlin (1999) state that the perceived impact that a technology will have on users is one of the factors influencing acceptance of the technology. For example, a technology that is likely to improve users' skill levels (e.g. a decision-making tool) is much more likely to be accepted than one which aims to reduce the need for users' skills (e.g. automating a key work function). Liker and Sindi (1997) found that perceived impact on skills valued by the user had a direct effect on intentions to use the new technology. Leonard-Barton and Kraus (1985) have also highlighted the impact of a technology on the user as being a key reason for resistance, stating that the two most common reasons for resistance to new technology are the fear of losing skills or power, and lack of personal benefit. One example of a situation where valued skills were lost is provided by Leonard-Barton and Kraus (1985). In this situation, supervisors who had excellent knowledge about an old system found themselves supervising subordinates who knew more about the new system than they did, causing a sense of losing control.

Issues around loss of control and loss of power by users have also been highlighted by Lapointe and Rivard (2005 – Case 1) who investigated the implementation of a new computer system in a community hospital. Despite a committee of physicians, nurses and other professionals choosing the software program and physicians being part of the process, major conflicts arose between physicians and nurses and then physicians and administration. Physicians initially complained that using the computer system was time-consuming and since they were paid by procedure, the extra time was costing them money. They also commented that completing clerical reports was not appropriate for a doctor and thus they saw this work as a threat to their professional status (Lapointe and Rivard, 2005). When the second stage was introduced requiring doctors to enter prescriptions into the computer, nurses refused to follow doctors' verbal directions and insisted that they use the system, indicating a shift in the previous power distribution (Lapointe and Rivard, 2005).

Introducing new technology then can have a profound effect on the way that the organisation is structured, potentially changing accepted work practices and power relationships. It is important to attempt to understand the nature of this impact before a technology is implemented and to develop an effective change management process to deal with the issues. It is also important to consult the users about the new technology early in the process since employees who have changes 'sprung' on them may have a negative 'knee-jerk' reaction until they can assess the impact of that change on their job role (Greenberg, 2005).

Joshi (1991) has developed an equity-implementation model, based on equity theory, to explain resistance to change (including the introduction of technology). In equity theory, individuals look at their input, the outcomes and the fairness in any relationship in which there is an exchange and compare themselves with others to assess whether their gains are equal to others. If they perceive that the outcome is not worth the increased input or that there is inequity when compared to others, they are likely to resist change but a perceived increase in equity is likely to lead to acceptance to change (Joshi, 1991). According to Joshi's equity-implementation model, users proceed through three levels of analysis when assessing a change in equity arising from implementation of change or a new technology. At the first level of analysis, users evaluate the benefits to be gained by the new system, such as better working conditions, and then evaluate the possible costs, such as increased workload. If they consider that the benefits outweigh the costs, they are likely to accept the new system, but if the costs outweigh the benefits, they are likely to reject the new system. At the second level of analysis, users will expect to reap some of the benefits that the organisation will gain from the new system, thus they will compare the changes in their own outcomes with that of their employers. The perception that the employer has gained relatively more from the change than the user is likely to result in the user taking an unfavourable view of the change. At the third level of analysis, users will compare themselves with other users to see if the new system benefits everyone equally. If they perceive inequity, they are likely to have a negative view of the new system. Overall, if users perceive the new system and its consequences as being equitable, they are likely to accept its implementation (Joshi, 1991).

Other researchers suggest that individuals focus more on how the technology will affect them personally. Leonard-Barton and Kraus (1985) argue that, in order for an operator to want to use a new technology there must be some personal benefit which must outweigh any personal costs such as the need to learn a new skill. This view is supported by Canton, de Groot and Nahuis (1995) who suggest that workers will embrace new technology if the benefits outweigh the costs. For this reason, younger workers are more likely than older workers to embrace new technology because the benefits would be longer-lasting for them and so would be more likely to outweigh the costs (Canton et al., 1995). Another potential difference between older and younger workers can be seen in a study conducted by Liker and Sindi (1997) who found a significant relationship between age and perceived impact of system use on valued skills. Older participants believed the new computer system would erode their valued skills or make no change, while younger participants believed the system would help to augment their skills and develop new ones. Canton et al. (1995) suggest that resistance will occur whenever the costs of introducing a new technology are borne by the workers. An example of this is labour-saving technology which usually results in loss of jobs. Similarly, workers who have particular skills with existing technology may resist the introduction of new technology as they will no longer have an advantage over other workers (Canton et al., 1995).

Individual Characteristics

In addition to the perceived usefulness, ease of use and impact that a technology will have on a user, the users' individual characteristics will also influence their likelihood of accepting a new technology. Greenberg (2005) suggests five key individual factors that make people resistant to change. The first factor is insecurity regarding loss of their job or reduction in pay, as changes in the workplace may be seen as a potential threat to job security. The second factor is fear of the unknown because change places people in unfamiliar situations where they may be uncertain about what is expected from them and whether their relationships with others will also change. The third factor is threats to the current social relationships within the organisation. Changes to the nature of the job may change relationships with work mates which could undermine friendship groups. The fourth factor is the need to break habits and learn new skills which provides challenges and requires effort. The final factor is a failure to recognise the need for change. Other factors highlighted by Wanberg and Banas (2000) include resilience (i.e. the ability to deal with adversity), information about the change, and change self-efficacy (i.e. the belief in being able to continue to function effectively despite change).

Other individual characteristics have been highlighted by Martinko, Henry and Zmud (1996) who identified numerous factors that can influence an individual's attributions, expectations and behaviours regarding information technology. Causal attributions relate to individuals' beliefs about the cause of failure which can be stable (constant over time) or unstable (changing across situations) and internal (within the individual's control) or external (outside the individual's control). The various combinations of these four elements are likely to lead to different outcomes regarding intention to use the technology, expectations of success or failure and affect toward the system. For example, if anticipation of outcome is negative and attributed to ability (internal and stable), the individual is likely to expect failure and therefore unlikely to willingly use the information technology (Martinko et al., 1996). However, if the anticipated outcome is positive and attributed to user effort, the individual is likely to expect success and therefore likely to use the information technology (Martinko et al., 1996).

Expectations regarding efficacy and outcome may also influence acceptance of a new technology such that individuals who expect to be able to perform well with the new technology and expect positive outcomes from the technology are more likely to accept it (Martinko et al., 1996). Individuals may also be influenced by co-worker and supervisor attributions, expectancies and behaviours (Martinko et al., 1996). If, for example, a co-worker or supervisor has difficulty with a new item of technology and blames their inability on the technology itself, other workers are also likely to attribute their difficulty or inability to the new technology. Also highlighted by Martinko et al. (1996) are technology characteristics such as perceived usefulness and perceived ease of use, which were discussed earlier.

Management support is another important factor and includes employee involvement in the implementation process. Autocratic management is likely to lead to resistance whereas supportive management is more likely to have a positive impact on users' perceptions of the usefulness of the technology (Martinko et al., 1996). Depending on past experiences with information technology, individuals may anticipate positive outcomes with the implementation of new information technology (e.g. ease of use, enjoying performing tasks), or negative outcomes (e.g. longer task completion times, loss of job) (Martinko et al., 1996). All the factors noted above can influence an individual's causal attributions regarding new information technology (Martinko et al., 1996).

Martinko et al. (1996) have categorised reactions to information technologies as either behavioural or affective. Behavioural reactions will be acceptance, rejection or reactance. Acceptance stems from prior positive experience and the user's belief in their ability. Resistance stems from the belief that outcomes will be negative and that the user does not have the ability to perform as required. Reactance will occur when outcomes do not match expectations and may involve an increase in effort to meet outcome expectations, or else may result in an individual lowering their expectations or changing their view of the appeal of the outcome. Affective reactions on the other hand include stress, apprehension and anxiety. These reactions are influenced by the user's positive and negative expectations about the technology. For example, positive expectations will result in positive affective reactions, such as excitement, anticipation and happiness with the system (Martinko et al., 1996).

In relation to perceived ease of use, Nov and Ye (2008) have argued that perceived ease of use may be influenced by a resistance-to-change personality trait. In a study on the acceptance and use of a university digital library, Nov and Ye (2008) investigated the relationship between the resistance-to-change personality trait and perceived ease of use. The relationships between perceived ease of use and computer self-efficacy, computer anxiety, screen design and the relevance of the technology for carrying out their tasks were also investigated. They found that when resistance to change and computer anxiety were high, perceived ease of use was low, and when computer self-efficacy, screen design and relevance were high, so too was perceived ease of use. In another study involving the use of a digital library by a group of university students, Hong, Thong, Wong and Tam (2002) also found support for a relationship between perceived ease of use and computer self-efficacy such that those with high levels of confidence in using computers were more likely to perceive the digital library as being easy to use.

The personal characteristics of individual people who use the system are therefore important to consider as drivers for resistance to technology. It may be the case that individual concerns about a technology are more imagined than real and adequate exposure to the technology may resolve some of these problems. This may in fact be what is driving the observations of Hong et al. (2002) in relation to ease of use and computer self-efficacy and computer anxiety outlined above.

Social Influences

Leonardi (2008) argues that the process of introducing a new technology is a social one with users being exposed to and influenced by other people's opinions, beliefs and stories about that technology. A strong social influence was observed by Hu, Clark and Ma (2003) in their investigation of resistance to a PowerPoint software program for use in Hong Kong public schools. Hu et al. found that teachers' opinions of PowerPoint were influenced by the opinions held by other teachers before but not after a training session. This suggests that once the teachers had experienced the program for themselves, they became more open-minded about its benefits (Hu et al., 2003). In another study of teachers' use of computers, Zhao and Frank (2003) found that pressure from other teachers increased the likelihood of a teacher using computers for their own purposes, and having assistance from other teachers increased the likelihood of teachers using computers with their students.

Although social influence appears to play a role in acceptance or resistance to technology, this is not always the case. For example, Davis et al. (1989) found no evidence of social norms having any effect on MBA students' behavioural intention to use a new computer system. Leonardi (2008) has also argued that social influence does not fully explain the acceptance or rejection of a new technology. In an investigation of the implementation of new software technology for use by performance engineers in the crashworthiness engineering division of an automobile manufacturer, Leonardi (2008) found that acceptance of the new technology was largely influenced by its perceived usefulness and reliability, with social interaction having only a minor influence.

Social influence therefore may have an impact on how a technology is perceived by the users depending on the characteristics of the social group of which the users are a part. Social influence seems to be particularly likely if the users have limited experience with using the technology. It seems that the more experience the users have with a new technology the less likely they are to be influenced by social factors. This is discussed further in 'Case Study of Resistance to Technology in Train Drivers' below, which describes a case study of resistance to technology in a rail setting.

Organisational Factors

The users of a new technology may also be resistant to the introduction of a new technology because they have negative feelings towards the organisation or managers who are in favour of the change (Dewan, Lorenzi and Zheng, 2004). Jian (2007) argues that resistant behaviours arise from organisational tensions which come into play when adopting new technology. Greenberg (2005) suggests four key organisational factors with regard to resistance to change. The first is structural inertia which relates to the fact that jobs are generally designed to have stability, with workers being trained to perform certain jobs and paid for the skills required for that job (i.e. change threatens this stable relationship). The second organisational factor is work group inertia where group social norms create pressure for work to be performed a certain way and changes may disrupt these expectations about the work. The third factor is threats to the existing balance of power that may arise from staff restructuring or promotions, potentially resulting in a loss of power or status (as discussed in 'Perceived Personal Impact' earlier in this chapter). Finally, previously unsuccessful attempts at change are likely to lead to worker doubts regarding any further change initiatives.

Employees' previous experience with the process used to implement previous technologies will shape whether those employees are likely to resist the implementation of a new technology. Previous negative experience with the introduction of new technology was found to be a major cause of resistance by social service caseworkers to the introduction of laptop computers in the field (Stam, Stanton and Guzman, 2004). This was exacerbated by an organisational climate in which the caseworkers felt that they were having the new technology forced upon them, as happened previously with the introduction of mobile phones (Stam et al., 2004). Caseworkers argued that mobile phones were simply a way of administration maintaining control over them at all times.

The organisational culture and the current position of an organisation will shape how easily a new technology is accepted by the employees (Dewan et al., 2004). For example, introducing a technology into an expanding company will be regarded more favourably than new technology aimed at downsizing in an organisation rife with conflict between unions and management (Liker et al., 1999). Likewise, introducing a new technology into an organisation with good communication and mutual respect between management and employees is much more likely to be accepted than in an organisation where employees have little or no respect for management.

The current relationship between the employees and management therefore appears to have an important influence on whether the employees will be resistant to the introduction of new technology. The organisational setting thus provides a context for resistance to technology and may well shape other reasons for resistance. For example, the employees' distrust of an organisation may lead to a negative evaluation of the impact of technology on their work, may fuel fears about their work (individual characteristics) and may shape negative social views which influence how the employees perceive the technology. This demonstrates the important point that while the different reasons for users resisting technology are separated out here for convenience of presentation, in reality, it is likely that the different reasons influence and mutually interact with each other. A summary of these reasons is presented in the box below. The way in which the organisation shapes resistance to technology is also discussed below in the context of the different forms that resistance to technology can take.

Why do people resist new technology?

- Technology not perceived as useful
- Technology not easy to use
- Perceived negative impact on skills, power/control
- Perception of benefit/costs
- Inequity
- Individual characteristics – uncertainty re job security; fear of unknown; threat to social relationships; not recognising need to change; attributions; expectations; past experiences; resistance-to-change personality trait
- Social influences
- Organisational factors.

What Are the Different Types of Resistance to Technology

Resistance to technology can take numerous different forms and has been expressed through relatively minor actions such as pointing out the flaws in a new technology and only partial use, to more serious actions such as refusing to use the technology and sabotage. Resistance to technology is usually a complex phenomenon that takes place in the social and organisational stream of events in which people work. The social and organisational context and the nature of the technology itself determine the opportunities for people to show resistance and shape how this is viewed by management, which may then lead to more extreme acts of resistance. The following sections outline the main forms that resistance to technology has taken in the literature, together with some examples.

Resisting Through Raising Concerns about Quality

A common way of resisting a new technology is to argue that the technology will impair the quality of the product or service that the organisation provides. A good example of this type of resistance is found in a study by Stam et al. (2004) that examined the introduction of laptop computers into an organisation providing social services. Many of the users (who were caseworkers) raised concerns about how the technology might compromise the quality of the services they provided to their elderly clients, something they valued highly. In contrast, other caseworkers were positive about the laptops, stating that it would free up more time for them to focus on their clients and therefore improve their services. The caseworkers who argued against the use of laptops stated that their clients were elderly and would feel intimidated by computers and distanced from the caseworker because the computer would be between them. Stam et al. (2004) found that, in practice, there was no basis for the caseworkers' concerns and given an aversion to computers in general by many of the caseworkers, suggested that the caseworkers' negative reactions were a shared strategy for resisting the technology, rather than an accurate portrayal of their clients' responses to the technology.

Asking Multiple Questions and Pointing Out Flaws

Asking multiple questions and pointing out flaws can be a form of resistance to new technology. In an investigation of resistance to a new computer system implemented in a health maintenance organisation, Prasad and Prasad (2000) found that employees asked multiple questions and made comments during training sessions about the possible negative health effects of constant computer use and pointed out flaws in the system. This form of resistance is subtle but was reported by employees as a deliberate attempt to resist the technology, with employees later stating that they had raised these questions to indicate to management that they would not passively accept management initiatives (Prasad and Prasad, 2000). Another example of this type of resistance can be found in continuous requests to redesign the technology. According to Smith and Douglas (1998), air traffic controllers involved in the design process of new technologies tend to continually request redesigns. While this could be interpreted as a genuine reflection of individual preferences, this could also be interpreted as resistance to the new technology and a way of putting off implementation (Smith and Douglas, 1998). While this form of resistance to technology is relatively minor, it is important to be aware of such resistance because in some cases resistance can escalate if not dealt with appropriately.

Not Using the Technology Correctly

Employees can resist the introduction of a new technology by deliberately using it incorrectly or only partially using it. A study by Jian (2007) examined the resistance of users to an enterprise-wide call-tracking software program implemented in a technology service organisation. The users were concerned about degradation of customer service, reduced social interaction between employees and the possibility of the software being a form of 'Big Brother'. Jian (2007) found that the users used the technology in a way that it was not designed to be used, used it for some customers and colleagues but not others, and in some cases refused to use the program or withdrew from using it if they had started using it.

Similarly, the study by Prasad and Prasad (2000) into resistance to a new computer system, discussed earlier, also found that some employees partially used the new system but primarily worked using the old manual system. While some of those employees did so because they felt more comfortable progressing gradually to a new system, others reported doing so as an act of resistance (Prasad and Prasad, 2000). Prasad and Prasad also found a slightly different form of this type of resistance where the employees accepted management's claims that the computer was very smart but then relied too heavily on the computer and less on their own ability and judgement which resulted in a deterioration in some areas of their work.

Pretending to Comply

One way that technology can be resisted is by pretending to be compliant. Pretend compliance is likely to take many different forms and an example of one such behaviour was observed by Mahoney (2010) when evaluating a new technology. Mahoney (2010) studied the evaluation of a new automated telephone monitoring system designed to address a shortage of caseworkers working with the elderly. In order to test the system, collaboration with a home care agency was required. The administrators of one home care agency agreed to participate but after nine months, the system evaluators had been unable to recruit any eligible elderly participants. Another home care agency was approached and agreed to participate but after three months, again no eligible elderly participants were found. When Mahoney investigated this lack of recruitment, she discovered that the union had rejected the project because of a perceived threat to the jobs of the case managers. Despite claims made during project presentations that the system would be of assistance in managing heavy caseloads, the union representatives were concerned that the new system would replace the case managers (Mahoney, 2010). Due to this concern, the union had instructed case managers of the home care agency to refer only those elderly clients who they knew would be ineligible in order to give the impression that they were cooperating (Mahoney, 2010). This example shows how people can pretend to comply and also shows the importance of including all relevant stakeholders in the process of developing and implementing a new technology.

Apathy, Disinterest and Refusal to Use the Technology

More serious resistance to technology is apathy or disinterest in the technology and, ultimately, refusal to use it. The case study by Lapointe and Rivard (2005), discussed above in 'Perceived Personal Impact', provides a good example of the potential seriousness of apathy and disinterestedness. The physicians at the hospital were unhappy with the introduction of the new computer system because they thought it would be time-consuming and as they were paid per procedure, they thought it would cost them money (Lapointe and Rivard, 2005). Resistance began with general apathy and disinterestedness which progressed to refusal to use the system. After a while, the physicians came to an agreement to take a stance together and made a formal complaint to the general manager, insisting that the new system be withdrawn. The general manager refused to withdraw the new system, which led to several physicians threatening to resign from the hospital. The breakdown in relations between the physicians and the management of the hospital caused by the introduction of the new technology ultimately led to six physicians being banned from the hospital, several more resigning and closure of the emergency room.

In another hospital case study conducted by Lapointe and Rivard (2005 – Case 3), a new computerised system was initially welcomed by physicians but after using the system for a while, they eventually felt that it would increase time for certain tasks and reduce their income (Lapointe and Rivard, 2005), a concern that would come under perceived personal impact (above). The physicians also argued that the system was a danger to patients and appointed a representative to discuss their concerns with the chief executive officer. As the resistance escalated, physicians felt that the new technology was a symbol of their loss of power in the organisation (again, perceived personal impact). When the physicians' representative issued an ultimatum, the hospital administration refused to allocate beds to the resistant physicians but the non-resistant physicians supported their colleagues, causing significant financial problems for the hospital and the system was finally withdrawn from those units.

Sabotage

Perhaps the most extreme form of resistance to technology is sabotage. In the study conducted by Prasad and Prasad (2000), discussed earlier in 'Asking Multiple Questions and Pointing Out Flaws', the more minor forms of resistance to technology escalated from questions and negative comments about the system to incorrect use of the technology to direct acts of sabotage. Prasad and Prasad were told of instances where employees had left drinks next to keyboards where they could be knocked over. Although no employees admitted to these deliberate acts, the opinions of most employees and management was that they were an intentional form of resistance. One employee was even considered to be a hero for her assumed part in one major sabotage incident in which new video display terminals were destroyed when the basement they were stored in was flooded (Prasad and Prasad, 2000). The water pipes in the basement had burst and although never proved, management and employees alike believed that the bursting of the pipes was not an accident (Prasad and Prasad, 2000). While much of the resistance reported by Prasad and Prasad was assumed rather than openly admitted, an act of sabotage by its nature is surreptitious and unlikely to be openly admitted. The fact that employees will consider sabotaging a piece of equipment is, in itself, a reflection of the extreme tensions that can be created in an organisation by the introduction of a new technology.

Conclusion

The case studies presented in this section show the different forms resistance to technology can take. While resistance is likely to be more common when employees have the technology imposed upon them with little or no opportunity to contribute to the design of the system, it seems that even when employees are invited to participate in the design process resistance can be found (Smith and Douglas, 1998). This shows that it is important to take the views of the employees seriously and to address concerns at an early stage. In addition to the views of employees, it is also important to consider the opinions of other key stakeholders in the development and introduction of new technology. The case study by Mahoney (2010) shows some of the problems that can result from not convincing a key stakeholder group (in this case the union) about the benefits of a new technology.

These case studies show how extreme acts of resistance can result from earlier more minor acts of resistance. In the case of the health maintenance organisation (Prasad and Prasad, 2000), questioning and pointing out flaws led to using the technology to degrade performance and acts of sabotage, and in the case of the physicians in the hospital (Lapointe and Rivard, 2005), individual acts of apathy and disinterest escalated to group refusal to use the technology. Lapointe and Rivard (2005 – Case 3) make the point that the resistance of physicians in the hospital moved from passive (apathy) to aggressive (refusal) in line with an increase in the level of perceived threat. As the resistance escalated, the perceived threats represented by the technology moved from an individual level, such as financial well-being, to a group level, such as loss of power (Lapointe and Rivard, 2005). The focus of the resistance also changed from the system itself in the initial stages, then to the significance of the system and finally to the system advocates (Lapointe and Rivard, 2005). In the early stages, resistance was independent of others and attitudes varied but in the later stages, physicians formed coalitions and resistance was planned (Lapointe and Rivard, 2005). Even physicians who were happy with the new system joined in the group resistance to support their colleagues as the resistance escalated.

It is clear then that a new technology is always introduced into a social and organisational context. In the section 'Asking Multiple Questions and Pointing Out Flaws' for example, the employees made the point that they engaged in this resistant behaviour to demonstrate that they would not passively accept management initiatives. As discussed earlier, resistance to technology may be as much a product of management/employee relations as about any real concerns with the introduction of the technology itself. The way that management respond to such resistance is also important since an unwillingness to listen to employee concerns can lead to further, more serious acts of resistance and a serious breakdown in relations between management and employees. This can ultimately lead to the withdrawal of the technology (which may have been expensive to develop and may have had considerable benefits for the employees and organisation) and in some cases a serious disruption to the business. For example, the physicians' resistance to the introduction of computers (examined by Lapointe and Rivard, 2005 – Case 1) led to the closure of the emergency room. This shows the importance of understanding the social and organisational culture into which a technology will be introduced and developing a detailed understanding of the issues being faced.

So far, all the examples of resistance to technology (see box below) have been conducted in the context of non-transportation domains. To show that these issues also occur in rail specific contexts, a case study conducted by the authors in an organisation where resistance to a new technology was encountered will be discussed in the next section.

How does resistance manifest itself?

- Raising concerns about quality
- Asking multiple questions
- Pointing out flaws
- Incorrect use of the technology
- Feigned compliance
- Apathy, disinterest and refusal
- Sabotage.

Case Study of Resistance to Technology in Train Drivers

Background

Resistance to technology has been extensively researched in domains such as education, medicine and sales but has received little attention in the rail industry or the transportation sector as a whole. Since these domains are quite different to rail, it is useful to show that the research and theories from these domains also apply to a rail context. To this end, a case study will be presented where resistance to technology has been observed in train drivers.

Resistance to technology was a topic that arose in the context of discussions about the introduction of new technology with an Australian long-haul freight organisation. During discussions with managers, we were told about a technology that had received extremely negative reactions after having been recently introduced into two train cabs. This technology was based on a decision support interface used in simulators. These negative reactions were reported by managers and drivers in subsequent focus groups and ranged from strong criticisms to acts of sabotage (e.g. cutting wires and spray-painting the screen). The technology had been developed and implemented with minimal consultation with users and had appeared in the train cabs with little or no explanation about its presence. Subsequent meetings with the developers of the system revealed that they felt there was little benefit to drivers in the context in which it had been implemented.

The meetings with groups of drivers revealed that some were passionately against the introduction of the interface. In addition to the acts of sabotage and refusal by shunters to use the locomotives containing the technology, drivers argued that it was a distraction and didn't capture their expertise. One driver was quite angry about the organisation trying to 'dumb them down' and was quite vocal in his condemnation of the interface. While a few drivers suggested that it would be good for training in the simulator, they stressed that it should not be used in the train cab. During these meetings it was revealed that, while the drivers had extremely negative views about the technology, few drivers had actually used it. These negative views had therefore not arisen from experience with the interface but were based on social influence, that is, drivers' opinions about the technology had been formed based on other drivers and supervisors talking negatively about the technology. As discussed earlier, the acceptance or rejection of a technology is at least partially socially driven and can be influenced by other people (Hu et al., 2003; Leonardi, 2008). In order to further investigate this resistance, a series of semi-structured interviews were conducted with drivers in the organisation where resistance was encountered.

Method

Three researchers conducted 62 one-on-one interviews with train drivers in two different depots within the same rail organisation. One of these depots was the same as the one in which the initial driver group discussions had been held. The interviews at this depot were held two years after the initial group discussions. There were 57 male and 5 female participants between 24 and 65 years of age (average 44.8 years) who had between 8 months and 37 years of experience (average 12.9 years). Interviews were recorded and later transcribed verbatim.

The researchers used a set of questions as a guide only, allowing the drivers to talk freely and following up any interesting responses. A sample of the questions used is shown in Table 3.1. (In order to ensure confidentiality, the name of the interface has been replaced with [interface].)

The participants were first asked if they had heard of the interface and those who had not were shown a picture of it and the aspects of the interface were explained. Only two locomotives had the interface installed in them approximately four years before the original group discussions with drivers had taken place. Thus it was anticipated that many drivers would not have seen the interface or would not have had the opportunity to use it prior to the interview. Those who had used the interface were asked questions relating to their experience with it, while those who had not used it were asked to consider the implications of having it installed in cabs. Data were collected and analysed on the drivers' opinions of the usefulness of each of the elements of the interface as a tool for learning and for use in the cab, as a tool for improving performance, ease of use, potential to improve safety, accuracy of the information displayed (for those who had used it) and impact on difficulty of the task. Data were also collected on whether drivers had received training for using the interface, if they had talked to other drivers about the interface, their attitudes to technology in general and, lastly, their attitudes towards the organisation as a whole. The researchers did not ask a specific question regarding attitudes towards the organisation but many drivers made comments about the organisation without prompting.

Table 3.1   Sample of questions used in one-on-one interviews

Seen or used interface	Not seen nor used interface
Have you heard of the [interface]?	Have you heard of the [interface] …
Can you describe it, including what it does?	What information do you think would be useful?
Is it easy to use?	Do you think it would make your work harder or easier?
Do you think it helps you to drive more safely?	Do you think it could help you drive more safely?
Does the [interface] help you monitor buff and draft?	Do you think it could help you monitor buff and draft?
Do your co-workers talk about the [interface]? If so, what do they think about it?	How do you feel about new technology in the rail industry?
How much training did you get with the [interface]?	Do you think that new technologies could improve safety in the rail industry?

Results and Discussion

The opinions of drivers in the one on one interviews were more positive than those given during the driver group discussions. Table 3.2 summarises the main findings of the interviews.

Note that the numbers shown in Table 3.2 may not be reflected in the tables throughout this section as some drivers did not respond to all questions. For example, some drivers may have stated that the interface would be distracting but not stated their opinion regarding its usefulness. Also note that, of those who said they had not heard of or not used the interface, some may have used something similar in a simulator but not mentioned it.

Table 3.2   Summary of the main findings of interviews

		Used in locomotive (out of 27)	Used similar in simulator (out of 16)	Not used and/or not heard of (out of 19)	Total
Good for training and/or in simulator		8	12	12	32
Useful in cab for day to day driving		16	7	11	34
Good for 'tweaking' performance		11	10	10	31
Accurate		9	N/A	N/A	9
Potentially distracting		17	8	12	37
Effect on difficulty of task	Harder	1	3	4	8
	Easier	6	3	5	14
	Both or neither	13	7	4	24
	Not asked/ don't know	7	3	6	16
Attitude to technology	Positive	29	13	20	62
	Negative	0	0	0	0
Received training		4	N/A	N/A	4
Attitude to company	Positive	0	2	0	2
	Negative	8	3	1	12
Attitude of other drivers spoken to	Positive	3	1	0	4
	Negative	10	4	1	15
	Both/neutral	2	0	1	3

One of the key determinants of whether people will accept or be resistant to a new technology is how useful the system is perceived to be (Davis, 1986). Overall, 32 drivers said that one or more of the features on the interface could potentially be useful. Two features were identified as being the most useful: (1) it is not always possible to know the buff (force created by compression of couplings) and draft (force created by the drawing out of couplings) characteristics along the train, thus displaying current and predicted buff and draft information could help reduce these forces; and (2) the elevation and curvature display would be useful for 'setting-up' actions and generally planning ahead. However, there were conflicting opinions amongst drivers regarding the need for these displays. Many drivers argued that a fully qualified driver must know the route so well that they should not need a display to show them the track profile. Some also argued that they knew what the buff and draft was like throughout the train all the time and did not need a display to tell them. Several drivers also stated that the methodology taught to drivers is sufficient to ensure optimal driving performance.

As can be seen in Table 3.2, three quarters of those drivers who had used something similar to the interface in a simulator and approximately two thirds of those who had never used it suggested that it may be useful for training purposes. In contrast, less than one third of those who had actually used the interface in a locomotive suggested that it would be useful for training purposes. Note that there was not a direct question about the usefulness of the interface for training purposes but this was a common theme, thus some drivers may have agreed but did not express their thoughts. Many of the comments were positive about using the interface for training but not for day to day use, for example:

> Interviewee J13: I think it would be [useful] at the start. I think as you drove more and more, you've worked out what your train was doing, I don't think you'd look at it.

> Interviewee JG03: It can give the driver an, a trainee driver I suppose, an understanding of how the length of his train affects his in-train forces and the track profile affects your train forces but to utilise it on a day to day basis it adds too much of a distraction from the primary focus of what the driver should be doing, looking ahead not looking at the screen.

Interestingly, in response to a direct question about its usefulness for day to day driving in the locomotive, approximately half the drivers in each group suggested that one or more elements could be useful, thus perceived usefulness does not seem to have been affected by use or non-use of the interface. In the following discussion, perceived usefulness refers to day to day use in the train cab.

Perceived accuracy of the information being displayed on the interface did not appear to influence drivers' perceptions of its usefulness. Most of the drivers who had actually used the interface in the cab were asked if they believed it to be accurate. Of those, nine thought it was accurate, ten thought it was not, and four were not sure.

Drivers were asked about the influence the interface might have on their job. Overall, eight believed it would make their job harder, 14 easier, 24 a combination of both or neither and 16 were unsure or not asked (see Table 3.2). As can be seen in Table 3.3 below, seven of the eight drivers who believed it would make their job harder thought it would not be useful in the train cab and only one driver thought that any part of the interface might be useful (the gradient display). In contrast, 13 of the 14 drivers who said it would make their job easier were positive about the usefulness of the interface (one did not respond re usefulness). Those who said it would not be harder or easier, or both, were almost equally divided between positive and negative about the interface's usefulness. This supports the view of Liker et al. (1999) that perceived personal impact (in this case, making the job more difficult or easier) may influence acceptance of a new technology.

Table 3.3   Influence of perceived task impact on perceived usefulness

	Positive re usefulness	Negative re usefulness
Harder	1	7
Easier	13	0
Neutral or both	13	10

Every driver expressed a positive attitude towards technology in general, thus resistance to change or technology in general is not likely to have affected drivers' acceptance of or resistance to the interface. Likewise, training did not appear to have any influence on the drivers' perceived usefulness of the interface. Only four out of the 27 drivers who used the interface in the locomotive had received training, of whom two were positive and two negative about the interface (see Table 3.4). However, those drivers who had received training indicated that it was minimal, although most thought it was sufficient. Of those who did not receive training, 10 thought one or more elements of the interface would be useful in the cab and seven thought it would not be useful.

Table 3.4   Influence of training on perceived usefulness

Training	Positive re usefulness	Negative re usefulness
Yes	2	2
No	10	7

The evidence for other drivers' opinions influencing perceived usefulness of the interface was uncertain. Of the drivers who expressed an opinion on the usefulness of the interface in the train cab, 22 said they had spoken to other drivers, four of whom had heard positive comments, 15 had heard negative comments and three had heard both or neutral comments (see Table 3.5). Attitudes to the interface matched opinions of other drivers for half of the drivers (i.e. positive talk/positive attitude to interface and negative talk/negative attitude). Two drivers who heard neutral or a combination of positive and negative talk were positive about the interface, and one driver who heard neutral talk was negative about the interface. Two drivers who heard positive talk were negative about the interface, and six drivers who heard negative talk were positive about the interface.

Table 3.5   Influence of others' opinions on perceived usefulness

Opinions	Useful	Number of drivers
Positive	No	2
Negative	Yes	6
Both or neutral	Yes	2
Positive	Yes	2
Negative	No	9
Both or neutral	No	1

Attitudes to the organisation did not appear to influence drivers' opinions about the usefulness of the interface, although it should be noted that drivers were not specifically asked about their attitude to the organisation and thus only 14 drivers commented about the organisation. Of the 14 drivers who made comments, two were positive about the organisation and 12 were negative (see Table 3.6). Of these, one was positive about the organisation but negative about the usefulness of the interface, one was positive about the organisation and the interface, seven were negative about the company and the interface, and five were negative about the organisation but still positive about the interface.

Table 3.6   Influence of attitude to company on perceived usefulness

Attitude to company	Useful	Number of drivers
Positive	No	1
Positive	Yes	1
Negative	No	7
Negative	Yes	5

Some drivers thought information from the technology could be used by management to help drivers improve their train handling skills. For example:

> Interviewee IR11: It would be terrific if this information can be downloaded ... a driver logs on ... and then it records information for that driver through their shift and then when they've finished, they sign out and so that information can be downloaded. Then it goes through to the bosses ... and they come back and go 'Well, your fuel consumption and your spring and buff forces are excellent, can you write down your methodology for us?' And then use that in contrast to someone who's probably not so energy-efficient and explain to them the problems they have with their methodology and 'Here's one here, which is exceptional. Can you adapt your methodology to this one?'

However, other drivers expressed concern that management and supervisors would have access to print-outs or disks and use them against the drivers. For example:

> Interviewee J5: You'll probably find that, these sorts of things you put in there, they'll be used against us, by these clowns down here. They'll go 'Oh geez, you were rough here, you were rough there, blah blah blah' ... So they say, 'You've got to change your driving style'.

Despite these concerns, for some drivers at least, it appeared that if assurances were given that this would not happen, the drivers would be keen to make use of such a system. For example:

> Interviewee L14: I think the problem is that some guys might see it as, like I said, you know, as checking on them, you know, 'Big Brother' sort of thing. I think that's one of the issues. I think, as long as management make it quite clear that they're not using it as a big stick. I mean, if they pull that thing down and say 'we're going to do investigations' when something happens, and say 'we're going to use that against you as evidence', I think we're going to have some issues. But if they say, 'No, it's a pure training tool, we pull it down to do training with you and that's it. We've got other systems that are actually for accident investigation, we're not going to use it for that', I think it will be fine.

The concerns raised by drivers here are similar to the organisational issues identified earlier where workers felt that the introduction of technology was akin to 'Big Brother' watching over them and a way of management maintaining control over them at all times (Jian, 2007; Stam et al., 2004). It seems from this study that, if assurances are provided by management that data will not be used against a driver, users are more likely to accept the introduction of the technology. However, the provision of assurances may not be accepted by the users if they do not trust management.

More generally, there were many comments from highly experienced drivers to the effect that the organisation did not value their expertise and had treated drivers harshly over particular incidents that had occurred. These negative feelings towards management were more prevalent at the location at which the interface had been trialled. This is also the location where the initial driver group discussions were conducted, where drivers were generally very negative about the interface and where acts of sabotage had been reported. No one recalled any extreme acts of sabotage during the one-on-one interviews, however, there were some clear signs of resistance. Many drivers mentioned that turning off the interface and refusing to use it was very common amongst drivers, and one driver mentioned that the monitor had been taped over:

> Interviewee JG10: ... we actually had some that had been taped up. Actually got masking tape and taped it over.

An important reason why people may be resistant to a technology is if they perceive a legitimate safety or quality issue (the 'Is Resistance Always Unreasonable' section discusses this issue in more detail). When introducing a new technology into a train cab there is always the potential for distraction. The issue of distraction was a major concern of approximately half the drivers. They were not asked a direct question about distraction but despite that, more than half commented on this issue. For example:

> Interviewee L8: I think it'd be distracting ... especially at night-time. You've got the extra light in the cab. There'd be a lot of distractions due to people, you know, watching it instead of watching where they're going or playing with it.

> Interviewee J6: ... I think anything like that, you could find yourself being distracted somewhat by that and you could find yourself not doing things you should have done ... looking at that rather than actually keeping your mind on the job.

In total, 33 drivers said the interface would be a distraction, four said it could be and only one said they did not think it would be (24 did not mention distraction at all). This may reflect the organisation's safety focus, and the perception that distraction is a major cause of signals passed at danger (SPADs). There was some comment within the organisation that the issue of distraction was being raised by drivers in order to resist the technology, however, this does seem to be an example of a legitimate concern that needs to be addressed in the design or training. The next section considers the issue of legitimate concerns leading to resistance to a technology, what we refer to as positive resistance.

This case study has identified a number of reasons why train drivers are resistant to the introduction of a new technology that are similar to those identified in the literature (such as perceived usefulness and organisational issues). This provides support for the use of the literature on non-transportation domains to explore the issue of resistance to technology in the rail industry. It is also interesting that participants had more positive views about the technology in the interviews than in the group discussions. While it is difficult to draw firm conclusions about these differences given that they were conducted nearly two years apart, this finding would be consistent with social influence where participants are more resistant to the technology in groups rather than in an individual setting where they have had the technology explained to them. The results of this study also show that different people within a group can have different reasons for resisting technology. For example, some participants in the study were concerned with perceived usefulness, while others were concerned with organisational issues. This suggests that groups should not be considered to be homogenous and interventions that try to minimise resistance to technology should be sophisticated enough to deal with multiple and varied causes of resistance. It should be pointed out that the technology in question was not implemented particularly well, with little or no consultation with drivers and even the developers of the system had reservations about its use. This is not the ideal way to introduce new technology. The 'Minimising Resistance' section below provides some advice about how to effectively introduce a new technology so that extreme resistance is avoided, but first we ask the question – is resistance to technology always bad?

Is Resistance Always Unreasonable

We have spent the majority of the chapter so far highlighting reasons for resistance to technology, describing some of the forms that resistance can take, and discussing a case study conducted in an organisation where the technology was not introduced particularly effectively. In doing this we have skipped over an important question. Is resistance to technology always a bad thing? The answer to this question is a resounding 'No'.

Mabin, Forgeson and Green (2001) make the point that resistance to change is not necessarily bad and can, in fact, be a good thing. Resistance can sometimes be beneficial in highlighting potential, unforeseen negative consequences of a new technology. The second case study involving the introduction of an electronic medical records system into a university hospital described by Lapointe and Rivard (2005 – Case 2) provides an excellent example of resistance playing an important part in the removal of problematic technology. The resident physicians were initially in favour of the new system, however, when a pharmacy module was later introduced they were concerned that it would be a threat to patient safety and could reduce the quality of care. Physicians formed a coalition and demanded the withdrawal of the pharmacy module. Their concerns were taken into consideration and the pharmacy module was subsequently withdrawn. In this case, resistance was not only reasonable but led to the removal of a technology which could have impaired patient care.

Another example of positive resistance was observed by Kirkley and Stein (2004) who found that some nurses were resistant to using computer-based clinical information systems. Interviews with nurse leaders revealed that nurses who were reluctant to use the clinical information system were not resisting the technology itself but the addition of workload to an already overloaded schedule, which they felt could take time away from the more important task of caring for their patients. This is a legitimate concern as studies have shown that entering data on a computer may take more time than paper methods, although it has also been shown that the quality of data is much higher (Kirkley and Stein, 2004).

Resistance to technology is partially a discourse that is used with management to try to get them to change or remove the 'offending' technology. As a result of this partial discourse, resistance will contain rhetorically constructed arguments that seem to be reasonable objections. These objections are often based on quality of service or impaired safety (as shown by the examples described in 'Why Do People Resist New Technology'). If it does not contain such arguments then management will simply dismiss the employees' concerns out of hand. This means that most resistance to technology will contain arguments that appear to be reasonable objections. Employees who are resisting technology for other reasons can produce a number of seemingly valid concerns that may lead to extensive redesign of the technology (Smith and Douglas, 1998). It is important then to treat employee concerns with respect but to also understand the underlying drivers of resistance so that objections to the technology can be placed in context. The next section suggests some ways to do this as we turn to a consideration of how to minimise user resistance to technology.

Minimising Resistance

Perhaps the main thing that people want to know about resistance to technology is how to manage it so that it does not cause a disruption to the business or lead to the rejection of the technology. Implementing a new technology usually entails a change in an employee's work role and should be managed through a change management process. A number of practical guides to successful change management have been produced (such as Mento, Jones and Dirndorfer's, 2002, 12 steps, and Paton and McCalman's, 2000, six key factors). Rather than attempt to regurgitate the information contained in these texts, we have selected a few key issues from the change management literature that are important to consider in minimising resistance to technology. Readers interested in the change management process are directed to the original texts. Like other sections, the issues identified here are strongly related to each other, and the separation of these ideas into sections generally serves a presentational purpose rather than a conceptual one.

Involving Employees in the Process

The cornerstone of minimising resistance to a new technology is to include the employees in the change process. Ideally employees should be included in all phases of the development and implementation of a new technology, from initial design to evaluation and roll-out across the network. Any change in an organisation is more likely to be successful if the people who will be affected are active in the change process (Leonard-Barton and Kraus, 1985; Nilakant and Ramnarayan, 2006). The involvement of people in the change process gives them a sense of ownership, which leads to commitment to change. In an organisation studied by Leonard-Barton and Kraus (1985), employee representatives were involved in the design and implementation of a new computer system. As a result of this involvement, the introduction went smoothly and the new technology was embraced by employees to the extent that, when bugs were discovered in the system, employees were keen to debug 'their' system. In contrast, lack of employee involvement was found to be the cause of resistance to the implementation of a new computer system in a health maintenance organisation (Prasad and Prasad, 2000). One of the reasons why it is important to involve employees in the change process is because they know their jobs better than anyone else and would therefore have information highly relevant and necessary to ensure change is beneficial (Nilakant and Ramnarayan, 2006). There is also an ethical aspect to employee involvement in the change process and since change affects the employee, it is reasonable for them to be given the opportunity to have some input (Nilakant and Ramnarayan, 2006). Chapter 4 on user feedback and chapter 6 on Future Inquiry discuss ways of including participants in the process of designing and implementing new technology.

Choosing Participants in the Change Process

In many cases, it is not feasible for all employees of an organisation to participate in the change process. Under these circumstances, a selection of employees should be made, ensuring that those chosen are credible and well-respected within the organisation (Axelrod, 2000; Leonard-Barton and Kraus, 1985). These are the employees who will not only assist in making decisions relating to change but will also be the ones persuading others that change is good. Thus it is important for these people to be opinion leaders, well-liked within the work group, and have an affinity with the people they are persuading. Nilakant and Ramnarayan (2006) identify three types of people who are particularly important to recruit as participants in a change process: connectors, mavens and salesmen. Connectors are people who possess large networks, mavens are knowledge accumulators who have the knowledge and social skills to start social epidemics, and salesmen are people who are good at persuading others to a particular perspective. Enlisting these three types of people generates commitment to change and ensures that the message spreads rapidly through the organisation (Nilakant and Ramnarayan, 2006). It can also be beneficial to enlist the aid of experts (such as consultants and university researchers) who may not necessarily be familiar with workers but who the workers are likely to respect due to their neutrality, knowledge and experience (Cialdini, 2007).

Marketing

Leonard-Barton and Kraus (1985) suggest that the implementation of a new technology should be regarded as a marketing endeavour that involves research on user needs and preferences. This approach should lead to a better fit between user needs and the product, good preparation of the organisation for receiving the product, and ownership of the technology by users. In addition to research on user needs and preferences, it is also important to consider the ways to persuade different groups about the benefits of a technology. One way of marketing a new technology is by including users in the development and implementation phases, as outlined above. Involvement in the development and implementation of a product leads to a sense of ownership, which encourages people to advocate the benefits of the new technology with their co-workers. As Cialdini (2007) has commented, supervisors are more likely than the chief executive of the company to be successful in persuading workers that a change is positive. It may also be necessary to market the technology more widely to the work groups through information sessions and meetings to consider worker issues before workers are exposed to the new technology in training sessions.

Targeting Employees

In order to succeed in reaching change goals, Nilakant and Ramnarayan (2006) suggest that support is needed from at least 30 to 40 per cent of employees within an organisation, though in practice this number is probably closer to 20 per cent. In addition, there is likely to be around 20 per cent of employees actively opposed to change, with the remaining 60 per cent undecided. Davis and Songer (2008) suggest that identifying and working with the resistant individuals before implementing a new technology enables an organisation to modify its process to reduce resistance. However, Baum (2000) suggests that management target the undecided 60 per cent in an effort to persuade them to support the change. Rather than targeting one or other of these groups, it seems that a sensible strategy would be to identify and target both groups with slightly different marketing approaches. It is also necessary to explore the different reasons why employees may be resistant. As our case study of resistance in the rail industry shows, there may be different sub-groups of resistant employees who may need to be persuaded in different ways about the benefits of the new technology. It is important then to understand the different groups of employees, the number of resistant employees and the different drivers of their resistance.

Persuasion

Nilakant and Ramnarayan (2006) argue that employee self-interest is important to keep in mind when persuading people to accept change, and it may be necessary to change the mindset of resistant workers through negotiations and persuasion. Simply using a logical argument is not going to sway people who do not see any personal benefits. According to Cialdini (2007), people are more likely to respond to the threat of loss than the promise of gain, thus when trying to persuade employees to accept change, the positive aspects of the change should be explained along with the potential negative aspects of not changing. It is essential, however, that the explanation of potential loss should not be framed in such a way that it could be interpreted as a threat, which is likely to lead to further resistance (Cialdini, 2007). Different techniques to persuade others have been well-researched in social psychology and include using credible sources, authority, counter-arguments, social influence and other less scrupulous methods (Cialdini, 2007; Levine, 2003; Thouless, 1953). Two examples of persuasive techniques that are likely to lead to acceptance of a technology are getting employees to publicly make a commitment to change (Cialdini, 2007) and introducing the technology in gradual phases, such as getting commitment for a pilot programme before full implementation (Nilakant and Ramnarayan, 2006).

Training

Another important step in the implementation of a new technology is to consider the training that people will receive. Insufficient or inadequate training can lead to dissatisfaction with the new technology, which can then result in resistance. For example, when a new software technology was introduced into the crashworthiness engineering division of an automobile manufacturer, performance engineers were not given full training on the use of the new system and were left to work out many of the features themselves (Leonardi, 2008). They often asked other engineers (who knew little more than they did) about the features and how to use the software, which led to much misinformation and a general misunderstanding of the true purpose and best use of the technology (Leonardi, 2008). An important point that arose from this research was that the attitudes of the engineers to the software were created by their own interaction with it but their expectations of the purpose of the software were based on misinformation from other workers and lack of training. This was despite the fact that they were initially excited about the possibility presented by the new technology (Leonardi, 2008). With correct training, it is likely that the resistance could have been reduced or even avoided entirely.

Voluntary Use of Technology

One strategy to minimise employee resistance is to make the technology voluntary when it is first introduced. In a longitudinal study of users from several organisations, Hartwick and Barki (1994) found that when the use of an information system was voluntary, user participation during the design process resulted in the users perceiving that the system was important and relevant to them. This perception then led to usage of the system. In contrast, when use of the new system was mandatory, user participation during the design process did not significantly impact upon system usage. Further, when system use was mandatory, the negative opinions of others were more influential regarding actual system usage (Hartwick and Barki, 1994).

There is some evidence that, as people become more familiar with a technology, acceptance increases (Kirkley and Stein, 2004), which means that a technology that is initially introduced as voluntary can gradually become a central part of the work as users begin to perceive its benefits and negative feelings diminish. The gradual introduction of a technology uses a persuasion technique known as foot-in-the-door (Freedman and Fraser, 1966). The principle of this technique is that people are more likely to agree to a larger request if they have already made a small concession. Thus, making a new technology voluntary rather than compulsory when it is first introduced should be considered as a strategy to minimise potential resistance.

Summary

Since introducing a new technology usually represents a change to workers' roles, a change management process should be used to manage the change. This section picks out a few key issues from the literature on change management to make a number of points about how to minimise resistance to technology (see box below). The cornerstone of minimising resistance to a new technology is user participation in the development and implementation process. In addition to reducing possible resistance, user participation in the process ensures that technology is centred around the user rather than having the user adapt to the technology, which is more likely to realise the benefits of the technology. User participation also creates a group of workers who advocate for the technology within their work group. Because this group will become advocates for the technology, this group should be chosen carefully to comprise opinion leaders and people held in high esteem. More generally, marketing activities should be conducted to promote the benefits of the new technology and this marketing should target different groups with varying levels of resistance and different sub-groups within resistant groups who may be resistant for different reasons. To adequately convey the benefits of a new technology it may be necessary to use well-established techniques of persuasion. Finally, it is important to consider the training that people receive in the new technology, since this is often the first opportunity many employees will get to interact with the technology. Effective training in the technology is vital to ensure that users know how to use the product correctly in order to get the most out of it, which has been shown to increase acceptance and reduce resistance. Support should also be available at all times to ensure users do not become frustrated with the system.

How can resistance be minimised?
- Employee involvement in the process
- Choosing the right participants in the change process
- Marketing
- Targeting the right employees in the right way
- Persuasion
- Training
- Voluntary use rather than compulsory use.

Conclusions about Resistance to Technology

Technological advancements in the rail industry will no doubt continue well into the future as organisations seek to improve efficiency and safety and to reduce operating costs. The introduction of new technologies can be seen by employees in a negative way, particularly where the technology creates a fundamental change to the worker's job role. Employees can be resistant to the introduction of new technology for a number of reasons: (1) if they perceive that the technology has a significant impact on their working life without having a significant benefit, (2) if they perceive that the technology is not useful or is difficult to use, (3) because they are scared of the unknown or of learning new skills, (4) if their peers have a negative opinion of the system, and (5) because they distrust management. Within any group of people who are resistant to the introduction of technology and even potentially within a single individual, there may be a combination of different factors. Resistance to technology can also take many different forms, ranging from questioning the technology and pointing out flaws, to using the technology incorrectly, feigning compliance, apathy, refusal to use the technology and sabotage. Research in the rail industry conducted by the authors has found similar reasons and types of resistance to technology as seen in other domains. While resistance to technology is not always a bad thing it is important to manage employees' resistance. One of the cornerstones of mitigating resistance to technology is to involve employees in the process of developing and implementing a new technology. The inclusion of users in this process helps to design a technology that is less likely to possess human factors issues, can help to maximise the benefits of a technology and creates a group who will advocate the new technology in their work group. It is also important to understand the reasons why people are resistant to technology in an organisation and to develop an appropriate plan to persuade these employees of the benefits of the change. In this way, resistance to technology can be effectively managed and technologies that are introduced to enhance operations and safety are more likely to realise their true benefits.

Acknowledgements

The authors are grateful to the CRC for Rail Innovation (established and supported under the Australian Government's Cooperative Research Centres program) for the funding of this research. We are also grateful to the drivers who participated in our focus groups, and the managers and drivers who participated in the one-on-one interviews for our case study.

References

Axelrod, R.H. (2000). *Terms of Engagement: Changing the Way We Change Organizations.* San Francisco: Berrett-Koehler Publishers.

Baum, D. (2000). *Lightning in a Bottle: Proven Lessons for Leading Change.* Chicago, IL: Dearborn.

Canton, E.J.F., de Groot, H.L.F., and Nahuis, R. (1995). *Vested Interests and Resistance to Technology Adoption*. Tilburg, The Netherlands: Tilburg University, Centre for Economic Research.

Cialdini, R.B. (2007). *Influence: the Psychology of Persuasion*. New York: Harper Collins Publishers.

Davis, F.D. (1986). *A technology acceptance model for empirically testing new end-user information systems: theory and results*. (Doctoral thesis, Massachusetts Institute of Technology.) Retrieved from http://hdl.handle.net/1721.1/15192.

Davis, F.D., Bagozzi, R.P., and Warshaw, P.R. (1989). User acceptance of computer technology: a comparison of two theoretical models. *Management Science, 35,* 982–1003.

Davis, K.A., and Songer, A.D. (2008). Resistance to IT change in the AEC industry: an individual assessment tool. *ITcon, 0,* 56–68.

Dewan, N.A., Lorenzi, N.M., and Zheng, S. (2004). Overcoming resistance to new technology. *Behavioral Health Management, 24,* 28–32.

Freedman, J.L., and Fraser, S.C. (1966). Compliance without pressure: the foot-in-the-door technique. *Journal of Personality and Social Psychology, 4*(2), 195–202.

Greenberg, J. (2005). *Managing Behavior in Organizations (4th edn)*. Upper Saddle River, NJ: Pearson Prentice Hall.

Hartwick, J., and Barki, H. (1994). Explaining the role of user participation in information system use. *Management Science, 40,* 440–65.

Hong, W., Thong, J.Y.L., Wong, W., and Tam, K. (2002). Determinants of user acceptance of digital libraries: an empirical examination of individual differences and system characteristics. *Journal of Management Information Systems, 18,* 97–124.

Hu, P.J., Clark, T.H.K., and Ma, W.W. (2003). Examining technology acceptance by school teachers: a longitudinal study. *Information and Management, 41,* 227–41.

Jian, G. (2007). "Omega is a four-letter word": toward a tension-centered model of resistance to information and communication technologies. *Communication Monographs, 74,* 517–40.

Joshi, K. (1991). A model of users' perspective on change: the case of information systems technology implementation. *MIS Quarterly, 15*(2), 229–42.

Kirkley, D., and Stein, M. (2004). Nurses and clinical technology: sources of resistance and strategies for acceptance. *Nursing Economics, 22,* 216–22.

Lapointe, L., and Rivard, S. (2005). A multilevel model of resistance to information technology implementation. *MIS Quarterly, 29*(3), 461–91.

Leonard-Barton, D., and Kraus, W.A. (1985). Implementing new technology. *Harvard Business Review, 63,* 102–10.

Leonardi, P. (2008, May). *Reclaiming materiality: explaining interpretations of technology and resistance to organizational change*. Paper presented at the International Communication Association Conference, Montreal, Quebec, Canada. Retrieved from http://citation.allacademic.com/meta/p230495_index.html.

Levine, R.V. (2003). *The Power of Persuasion: How We Are Bought and Sold*. Hoboken, NJ: John Wiley & Sons, Inc.

Liker, J.K., Haddad, C.J., and Karlin, J. (1999). Perspectives on technology and work organization. *Annual Review of Sociology, 25*, 575–96.

Liker, J.K., and Sindi, A.A. (1997). User acceptance of expert systems: a test of the theory of reasoned action. *Journal of Engineering Technology Management, 14*, 147–73.

Mabin, V.J., Forgeson, S., and Green, L. (2001). Harnessing resistance: using the theory of constraints to assist change management. *Journal of European Industrial Training, 25*, 168–91.

Mahoney, D.F. (2010). An evidence-based adoption of technology model for remote monitoring of elders' daily activities. *Ageing International.* doi: 10.1007/s12126-010-9073-0.

Martinko, M.J., Henry, J.W., and Zmud, R.W. (1996). An attributional explanation of individual resistance to the introduction of information technologies in the workplace. *Behaviour & Information Technology, 15*, 313–30.

Mento, A.J., Jones, R.M., and Dirndorfer, W. (2002). A change management process: grounded in both theory and practice. *Journal of Change Management, 3*(1), 45–59.

Nilakant, V., and Ramnarayan, S. (2006). *Change Management: Altering Mindsets in a Global Context.* London: SAGE Publications.

Nov, O., and Ye, C. (2008). Users' personality and perceived ease of use of digital libraries: the case for resistance to change. *Journal of the American Society for Information Science and Technology, 59*, 845–51.

Paton, R.A., and McCalman, J. (2000). *Change Management: a Guide to Effective Implementation.* London: SAGE Publications.

Prasad, P., and Prasad, A. (2000). Stretching the iron cage: the constitution and implications of routine workplace resistance. *Organization Science, 11*, 387–403.

Robinson, L., Marshall, G.W., and Stamps, M.B. (2005). Sales force use of technology: antecedents to technology acceptance. *Journal of Business Research, 58*, 1623–31.

Sarter, N.B., Woods, D.D., and Billings, C.E. (1997). Automation surprises. In G. Salvendy (Ed.), *Handbook of Human Factors and Ergonomics (2nd edn)* (pp. 1–25). Hoboken, NJ: John Wiley & Sons, Inc.

Smith, A.L., and Douglas, A. (1998). Introducing new technology to the air traffic controller: implications for skill acquisition and training. *IEEE*, F54-1–F54-8.

Stam, K.R., Stanton, J.M., and Guzman, I.R. (2004). Employee resistance to digital information and information technology change in a social service agency: a membership category approach. *Journal of Digital Information, 5*(4). Retrieved from http://journals.tdl.org/jodi/index.php/jodi/article/view/150/148.

Thouless, R.H. (1953). *Straight and Crooked Thinking.* London: Pan Books.

Wanberg, C.R., and Banas, J.T. (2000). Predictors and outcomes of openness to changes in a reorganizing workplace. *Journal of Applied Psychology, 85*, 132–42.

Zhao, Y., and Frank, K.A. (2003). Factors affecting technology uses in schools: an ecological perspective. *American Educational Research Journal, 40*, 807–40.

Chapter 4

Effective User Feedback:
The Practical Value of Mock-ups

Gareth Hughes
Railcorp NSW, Sydney, Australia

Airdrie Long
Considered Solutions Pty Ltd, Sydney, Australia

Anne Maddock
Railcorp NSW, Sydney, Australia

Chris Bearman
Central Queensland University, Appleton Institute, Adelaide, Australia

Introduction

Chapter 3 provided a review of the potential for the resistance to technology to impact on the utility of new technologies within the rail industry. A number of reasons and motivations for encountering such resistance were identified and strategies for mitigating its effects were proposed. The chapter suggested that some initial resistance to technology could be a positive thing as it can potentially represent a healthy reality check on organisational intentions. Further, it was stated that 'the cornerstone of minimising resistance to a new technology is user participation in the development and implementation process'. In this chapter, we discuss our practical experience of obtaining user participation in the development and design process by focusing on the use of mock-ups within the project environment.

A mock-up can be defined as a physical representation (or model) of a technology (or parts of a technology) constructed to investigate aspects of the design. Mock-ups can range from a simple cardboard or foam representation of the design to high-tech working models. In some respects it can be argued that a mock-up is effectively a type of simulator. Chapters 8 and 9 deal with some of the issues around the use of simulators and the discussions in those chapters around specification and collection of data and issues of fidelity are also relevant to the use of mock-ups. We discuss some of these issues briefly here but the reader interested in a more in-depth discussion of these issues is directed to these other chapters. This chapter focuses on the benefits and challenges of using mock-ups within a design context, including a short discussion of some of the commercial implications. The chapter focuses primarily on projects that are concerned with the design and development of hardware. However, it can be argued that the principles identified are equally applicable to software development. Therefore some discussion of the software domain is also included.

There are two key factors that influence the effectiveness of mock-ups. The first of these factors concerns the validity of data that is gathered, which depends on the sample of users involved in the process, the quality of data collection and the reliability of the data analysis. The second relates to the physical characteristics of the mock-up with respect to the timing of its use within the project lifecycle. This chapter further explores and explains the relevant factors that influence the successful use of mock-ups from both of these perspectives. Ultimately, the chapter demonstrates that, because mock-ups are tangible, three-dimensional, realistic and understandable, when used appropriately they can provide a cost-effective and efficient method to involve end users in the design process and gain valuable design information.

Why Use Mock-ups in the Rail Industry

There is a vast body of literature, in virtually every operational domain, that supports the necessity for a user-centred approach to design and emphasises that a failure to adopt such an approach can be a significant source of risk (e.g. Buxton, 2007; Krug, 2006; Rail Safety and Standards Board, 2008; Wilson and Haines, 1997). This philosophy is also backed by the inclusion of requirements for user involvement in many standards, such as: ISO 6385 Section 3.1 (ISO, 2004) and AS 4024.1901 Section 5.1 (Standards Australia, 2006). Mock-ups are a useful way of achieving user involvement and are widespread within rail design projects.

The primary purpose of a mock-up is to obtain feedback from the future users of a system regarding their views on the acceptability and utility of the emerging design. It is relatively common practice within major fleet procurement projects to include the provision for the development of a high-fidelity mock-up as part of the contractual requirements and a number of studies have explored this process in detail (Baxter and Hughes, 2008; Long and Hughes, 2011; Wright, 2007). However, on many occasions within the rail industry user feedback is still not sought appropriately nor is it used effectively within the design process. This cannot be good for the industry since a lack of consultation sets the scene for the kinds of resistance described in chapter 3. The case study presented in chapter 3 is a particularly good example of the problems that can occur when a technology turns up in the train cab without any prior consultation with drivers.

The developers of technology can be reluctant to obtain user feedback because of the perception that the users may manipulate the process for their own ends, such as using the feedback process to resist technology (see chapter 3) or holding out for 'nice to have' but unnecessary features. Such manipulation of the feedback process can lead to significant time and cost overruns, a myriad of unnecessary features that increase maintenance and training requirements, or a new technology that is simply a like-for-like replacement of the old technology. In severe cases, user manipulation of the feedback process can lead to the organisation being unable to manage the change process and implement the technology. In order to avoid these undesirable outcomes it is essential that:

- there are structured processes for gaining user feedback and the framework for acceptance or rejection of suggested improvements is defined
- the organisation goals are clearly articulated to users and any areas of change that are not negotiable are clearly defined.

Careful management of the process of obtaining user feedback in conjunction with good management practice, a good change management process and an ongoing good relationship between management and staff will go some way towards reducing the problems associated with users manipulating the feedback process. However, it is clear that the benefits of engaging in the process (from a better user-centred design and in promoting acceptance) far outweigh the risks. Simply put therefore, from a human factors perspective, taking account of user requirements is a fundamental tenet of the design process.

Clearly there is a need to dispel negative perceptions regarding the use of user feedback within the industry. One way in which this can be achieved is through a better understanding of the effective use of mock-ups within the design process. Towards this aim, the rest of the chapter discusses when to use mock-ups and how to make the best use of them. Towards the end of the chapter we provide a number of case studies to demonstrate the practical use of mock-ups.

When to Use Mock-ups within the Design Process

The previous section described 'why' we should use mock-ups. This section describes 'when' to use mock-ups within the design process.

The availability of Rolling Stock, Trackside Equipment, Signalling and Control Systems and Rail Maintenance equipment has a significant influence on the ability of a rail operator to run a safe and reliable railway. The design and development of new or replacement equipment of this type is a significant undertaking for any organisation, and one which is often contracted out to third parties since many organisations no longer have the resources available to complete such an activity. Indeed, in many very large-scale projects the management of the procurement process itself is also contracted out. Thus, in today's design environment, there are often a number of complex relationships between the organisations involved as each organisation potentially brings their own objectives to the process. Therefore the overall context in which the design is being developed can significantly impact on the desire or ability to incorporate user feedback into the design process.

The actual definition of stages in the design process is dependent on the organisation's own internal procedures. For the purpose of this chapter a generic and simplified model is proposed as, in principle, most design processes involve the development of a concept design, a preliminary design and finally a detailed design. Each stage in this process is typically accompanied by a formal review undertaken by or on behalf of the purchaser. Successful acceptance at each stage is also typically linked to a contractual milestone and payment. In theory, the design should not progress until acceptance of each stage has been achieved. However in practice this can no longer be the case, as commercial pressure dictates the need for parallel development of the design and the manufacturing process.

It is possible for users to be involved in all stages of design and as a rule of thumb it is usually considered that the earlier users are involved, the better. However, there is a trade-off here because:

- when provided with too many options, users can find it difficult to articulate the differences between them
- if the design is too conceptual, users can find it difficult to relate their current experience to the concept being presented.

In the early stages of design only high-level information can be provided, often in the form of concept diagrams, preliminary engineering or CAD (computer aided design) drawings etc. As the design progresses, more detailed drawings and information can be presented. While there is value in

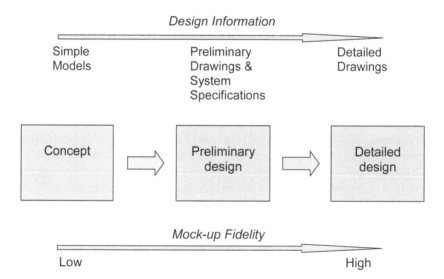

Figure 4.1   The relationship between the fidelity of mock-ups and the design process

obtaining user feedback during these periods it should be noted that users are not generally trained engineers and so their ability to fully comprehend and interpret detailed engineering drawings may be limited. Mock-ups, which enable designers to provide a three-dimensional representation of the system, play an important role in the process of helping users fully understand the emerging design. Clearly as the design process progresses the fidelity with which a three-dimensional model can be presented increases. Figure 4.1 illustrates the relationship between the level of fidelity of mock-ups and the design process. The use of mock-ups in each stage is discussed below.

Concept Stage

At the concept stage, one of the first data collection activities is typically to elicit user feedback about the existing system. There are many methods for doing this including individual and group feedback sessions which are generally used to gather experience of using the system, review of incident data which can be used to identify any potential shortfalls in the current system design and scenario analysis which involves the user talking or walking through specific tasks or operating scenarios. It is recognised that some organisations may be wary of using such data to influence the design process because sample sizes may be limited or there may be a lack of belief in its validity, particularly in those cases where the data gathered does not match the 'anticipated response'. However, it should be strongly emphasised that the output of such activities actually provides a very valuable source of data to the design process as it provides a rich contextual account of the issues and often represents the only way of truly characterising the current use of the system. Chapter 5 discusses in some detail the issues around collecting, analysing and using this kind of data.

Additionally, of course, it is not the case that all the information gathered should necessarily be accepted on face value. For example, an organisation may wish to conduct further investigations using different methods into any issue that could significantly impact the design prior to design decisions being made. Chapters 5, 7 and 9 discuss a range of qualitative and quantitative methods for investigating a particular issue in more detail.

Following initial data collection on existing systems and still during the concept stage, it is usually possible to identify the key areas of the design where there may be the highest risk of a lack of user acceptance. These areas could include the comfort of the user, concerns around safety, a requirement for increased productivity or where the new design includes significant novel features. At this stage the use of low-fidelity mock-ups can provide designers with useful information, such as the high-level relationships between equipment groups, the impact of changes of equipment layout on task sequences and user preference between alternate options. Mock-ups at this stage also help to begin the process of familiarising users with the emerging design.

Preliminary Design Stage

As the preliminary design develops, more detailed design information becomes available through system drawings, specifications and more recently through the application of computer generated models. It is advisable to include users in the review of such information regarding human interactions with the system. It should also be noted that this process should include consideration not only of major elements but should also involve consideration of interactions with equipment used for secondary tasks. In rolling stock procurement, for example, this means considering tasks such as shunting, local isolation of door controls, location of fire-fighting or passenger assistance equipment etc., in addition to the obvious focal area of the crew cab.

Historically, this information has been presented in a two-dimensional format and it is often the case that users, who are not used to working with this type of information, find it difficult to interpret from a final design perspective. Even with the advent of three-dimensional modelling, which is often inclusive of a manikin-type representation of a user, it is still likely that difficulties will be experienced. During a recent application of this type of process some users commented that they thought the manikins were unrepresentative, and that the colours used in the model did not give the look and feel of a real train. Additionally, due to the complexity of the software involved it is necessary for a technical designer to drive the software model. This can lead to the perception that the exercise is more of a demonstration process rather than one to gain real user feedback.

At this stage of the process mock-ups can add real practical value. Three-dimensional mock-ups can be used to investigate specific areas of concern arising from other reviews and to confirm design decisions that have been made as the design has evolved. At this point mock-ups can comprise basic plywood and block models with photographs or diagrams of displays and controls. They can be used to confirm the relationships between task components and can also form the initial basis for procedural reviews.

Detailed Design Stage

At the detailed design stage it is generally possible to replicate the actual intended design solution within a mock-up. In many cases the mock-up can include actual pieces of equipment and can facilitate reviews of some dynamic components of the design such as the force required to operate controls, the level of feedback to users about control positions, etc. Additionally, this level of mock-up often allows users to experience the actual look and feel of the design solution as the overall finish is often highly representative of the final design. In many respects this mock-up should be the way of fine-tuning the design for those components that it has not previously been possible to assess in this way.

How to Make the Most Effective Use of Mock-ups

This section addresses the issues of 'how' to make the use of mock-ups most effective and successful. As previously stated there are two things that significantly influence the effective and successful use of mock-ups. The first of these relates directly to the human factors issues concerning the sample of users involved and the validity of data that is gathered. The second relates to the actual characteristics of the mock-up and the timing of its use within the project lifecycle.

Human Factors Issues

There are a number of elements that must be considered in order to make the most effective use of mock-ups. First, the whole exercise must be properly planned. This involves ensuring that all the data requirements are identified in advance and all logistical arrangements are clearly articulated to users in a timely manner. Second, it is imperative that a representative sample of users is involved and that data collection and analysis methods are agreed. Finally data must be made available to those users who participated in the process.

Planning
The data requirements and the level of planning required can vary considerably depending on the stage within the design process and the type of mock-up to be used. Processes to gather data from high-fidelity mock-ups can cost many thousands of dollars due to the requirement to release operational staff from their duties and to have appropriate experts running the data gathering process. Therefore it is essential that this process provides value for money. An important consideration here is ensuring that the mock-up is located in a convenient location since this minimises the amount of travel time required by the participants.

On the other end of the scale some real benefits may be gained from the *ad hoc* use of very low-fidelity mock-ups. However, even in this case the advantage of having a structured and auditable process can be significant at later stages of the project should questions regarding early design decisions be raised again in the future.

It is essential to identify the actual data requirements that are needed at the end of the process since it is not only time-consuming but also highly embarrassing to find that a subsequent exercise needs to be carried out because the correct data were not gathered the first time. It is also equally important to make apparent to both users and the design team the limitations of the data that will be gathered and to also specify what data is not able to be collected given the fidelity of the mock-up proposed. Of course in reality the relationship between the data requirements and mock-up fidelity is an iterative process within the design team and one that needs to be resolved in a timely manner in order for the data collection exercise to be a success. This discussion is generally easier for those elements when there is a requirement to collect data regarding a specific design issue as opposed to a more generic requirement at the end of a design stage.

For example, in many mock-ups the representation of equipment and controls does not realistically replicate the forces required to operate them. Therefore, data cannot be collected in relation to this issue and a data requirement remains which can only be finally resolved through a higher fidelity mock-up or during the period of formal acceptance testing. The decision as to the requirement for a higher fidelity mock-up to review a specific item should be based on an assessment of the risk of having to change the item at a much later date in the design process or risk a lack of user acceptance. For example, if a control is specified as the same as the existing control then there is relatively little risk. However, if a new supplier or new design is involved then the level of risk is higher.

User Samples

Many rail organisations have formalised processes for user consultation and it is essential to ensure that when projects and mock-up reviews are established that the organisational processes for consultation are followed. Typically, however, this form of consultation is focused on the attendance of unions and union representatives within a formal consultation structure. For example, a crew cab committee in a rolling stock procurement project would be a typical mechanism of this type. This process can be invaluable, as the participants are often highly passionate about having their views heard. However, it should also be noted that this process could be subject to influence as part of a wider political agenda and the representation is usually limited to a small number of people who tend to be at the more experienced end of the user spectrum.

In addition to any formal consultation structure it is highly recommended that the user feedback process be extended to include a wider audience. In this way a wider range of users with different levels of experience can be represented and the organisation can have more confidence that the data gathered is a better reflection of the overall user population. While there is often a focus on experienced users it should be noted that the needs and feedback from inexperienced users is equally important within this process. The selection of the user populations should, as a minimum, take account of the user group, gender and level of experience. Depending on the actual context there may also be the need to account for other factors such as preference for driving in a standing or sitting position, left/right handedness, etc.

Of course there are also practical constraints on this process, such as the potential availability of users (who can often be limited in both number and opportunity). It is important to recognise therefore that the results of such exercises are not generally amenable to statistical evaluation and should be examined from a more qualitative perspective. Chapter 5 discusses qualitative methods in more detail.

Data Collection and Analysis

Finally, it is essential to develop and agree on a formal process for the data capture and analysis. The benefit of the discipline associated with this activity is that it forces the organisation to actually identify the real data requirements and therefore to further identify the most appropriate means of data capture. Robust and transparent processes are required in order to ensure that the data can be used and relied upon to feed into the consideration for the design, particularly in those cases where individual opinions on different options may differ.

It is necessary to gain agreement upfront for the data collection methods as part of the formal consultation process. This is especially the case where video or verbal recording of interviews or other data collection processes are required. Other tools that are useful are questionnaires and structured interview protocols although some of the more insightful comments often arise as a result of unprompted activity on behalf of the user.

A combination of individual and group based methods can often work well. In a recent example we used a process that began with a general group-based review and introduction to the system, followed by walk-throughs of tasks with individuals and ending with a group-based feedback session. Although this may seem inefficient as there were some periods when a specific individual was not involved in a review process, the overall approach allowed us to minimise the downtime for any individual by having more than one person carrying out data collection and to meet the requirement to roster staff off for a set minimum period.

Feedback

It is essential that users are given feedback about the results of the exercise in order to validate the results and to enable further similar exercises. Additionally it is very important to provide an explanation of what changes are going to be made and what changes are not going to be made following the exercise. An honest and transparent approach to giving this feedback is the best option. For example, if it is simply too costly to adopt a particular approach or suggestion then this should be the reason given for its rejection.

The Actual Mock-up

The actual characteristics of the mock-up and the timing of its use within the design lifecycle may have a significant impact on its effectiveness. 'Why Use Mock-ups in the Rail Industry' above described in general terms when mock-ups could be used in the design process. Experience has shown that one of the main problems has been a reluctance to utilise low-fidelity mock-ups early in the design process combined with a high reliance on high-fidelity mock-ups at a late stage.

There are a number of potential reasons for this reluctance. There is a view that the low level of fidelity of a mock-up will be interpreted by users as a lack of quality of the final product or alternatively the mock-up will not be good enough to raise comment. However, quite the opposite has been found, with users appearing to be more willing to comment freely and constructively on rougher mock-ups (Schumann, Strotthotte and Laser, 1996; Snyder, 2003). Snyder (2003) speculates that reviewers are more willing and provide more comment on rougher mock-ups because rougher mock-ups suggest the design is still on the drawing board while slick-looking mock-ups suggest a more finished, considered design. When only a high-fidelity mock-up is specified, concerns about the lack of quality of the final product can be more justified. There is often a reluctance to release the mock-up until it is of sufficient quality since this type of mock-up is often used for additional purposes such as public exposure or corporate publicity. In these cases organisations want to ensure that the finish of the mock-up is highly representative of the quality of the final product. The consequence of this delay to the delivery of the mock-up is that when it does actually become available it is no longer possible to influence many aspects of the final design.

Commercial Considerations

The effect of commercial pressures on the use of mock-ups is inherently linked to the type of mock-up being provided and its timing within the project lifecycle. In general terms the later the mock-up is provided the more costly it will be for changes to be made. In a commercial design and manufacture contract it is in the interest of both the contractor and the client organisation to provide a solution that meets both the required specifications and the needs of the actual users. Non-acceptance of a system by users often represents a significant source of commercial risk (as discussed in chapter 3). In cases where the specification can be sufficiently concrete and measured absolutely, there is usually relatively little room for disagreement, except where the designers make a request for a variation to the specification. In terms of user requirements however the situation is usually more flexible and this can lead to problems incorporating user feedback into the design process. From the designer's perspective it is understandable to view user feedback as a risk to project development and delivery, especially in those cases where unstructured processes for gaining appropriate user feedback are employed. In such cases it is not unusual for the user to change their mind halfway through the design process or to request alternative or additional features outside of the specification.

The perspective on user feedback can also change over the duration of a project. In the early days there is often a high level of enthusiasm within the project or design team for data collection and analysis to shape the emerging design. As the project progresses and the design becomes more fixed, there is an increasing reluctance to change, primarily due to the cost, effort involved and potential impact on project schedule. This effect can be so extreme that in some cases, by the time a high-fidelity mock-up is delivered, it is only possible to make relatively trivial changes. This is not such a problem if an appropriate use of lower fidelity mock-ups has been made throughout the process, since the user feedback will have been incorporated into the high-fidelity mock-up, which should then only require fine-tuning.

In summary, from a commercial perspective there are a number of key issues that must be considered within the contract specification, as follows:

- The proposed timing and fidelity of mock-ups
- The location of the mock-ups and the time that they will be available for use
- The process through which changes are made to the design based on the feedback obtained.

It is highly beneficial for the client organisation if these issues are also formally tied to contractual milestones.

Software Projects

Software projects also follow a structured design process and there is the same need to gain user involvement and feedback through the development of the product. In terms of low-fidelity simulation there are numerous technologies available that allow for the development of screen-based mock-ups of the user interface. These are low fidelity in that there is often no action as a result of a control input but can be highly representative of the proposed interface design.

In software projects it is common to employ usability tools as a means of capturing user feedback. These tools are often used in conjunction with a high-fidelity prototype of the software and typically comprise two elements. One element is a set of generic usability guidelines that can be applied by a non-user. For example a statement such as 'no more than three levels of hierarchy may be employed within the menu structure' could be used. The second element involves a structured elicitation process based on feedback from a potential user attempting to complete a set of specified tasks within the system. Case Study 4 illustrates the use of this kind of process.

Case Studies

This section presents four case studies that illustrate the use of mock-ups in the rail industry and serve to highlight the points made in previous sections of this chapter.

Case Study 1 – A Retrofit Project

A control was being added to an existing driver's workstation. The control had been proposed as a foot control but the control only needed to be pressed intermittently and there was a potential problem with limited foot space.

Figure 4.2   **Low-fidelity mock-up used to assess where drivers would and could place their feet while still easily pressing a foot control intermittently (from Long and Hughes, 2011)**

An initial rough and ready low-fidelity mock-up was used to examine the placement of the control system (Long and Hughes, 2011). The question was: Is there a location on the sloping footrest for the control such that drivers could comfortably place their feet on the footrest when not using the control while still easily reaching it when needed? (Long and Hughes, 2011). The footrest was replicated with a cardboard box cut to the right slope and a push-button (see Figure 4.2). The mock-up took half an hour to assemble and a further hour in the office to test with a range of people of different sizes (Long and Hughes, 2011).

To review the design a seat was provided and a number of people were asked to sit at the workstation. The sample of people included both novices and drivers. They were informed that they needed to rest their feet on the footrest and then occasionally press the button. Each person was asked whether they could find a comfortable position without pressing the button and whether they could press the button easily. The data gave the project team confidence that the concept was going to be successful despite the 'quick and dirty' nature of the mock-up (Long and Hughes, 2011). This case study presents an example of a very low-cost low-fidelity simulator providing valuable insight for the project team at a very early stage of the design process.

Case Study 2 – New Train Design 1

A new train was being procured. Purchasing a new train is an expensive exercise and the design needs to be carefully considered because trains typically have a design life of about 30 years.

The supplier's designers developed a foam core mock-up of the crew cab at the concept stage (Long and Hughes, 2011). The basic cab structure was made of wood and then covered in foam core board to represent the shape being proposed.

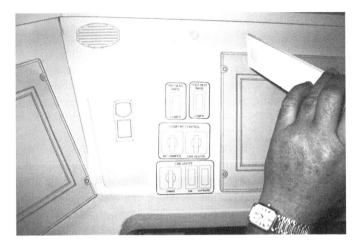

Figure 4.3 **Mock-up made from foam core board, hook and loop fasteners and photocopied controls and displays (from Long and Hughes, 2011)**

Photocopies of controls, displays etc. mounted on foam core board and attached by hook and loop fasteners were then attached at the appropriate places (see Figure 4.3). The designers involved estimated that the time required to prepare this mock-up was about a day.

The researchers observed train crews carrying out a number of crew tasks embedded in scenarios. As the crew were going to be using both the new train and the existing trains interchangeably it was important that crew expectations were considered and negative transfer opportunities were minimised. The supplier's designers, the client technical staff and crew discussed the experiences and learnt valuable information. Designers watched as the drivers instinctively went for the horn and the controller to stop the train (Long and Hughes, 2011). Also during the review a peculiar local practice was observed. On questioning it was found to be the result of sub-optimal design in the existing fleet. As a result the new train design was able to improve on this feature.

This case study also illustrates the potential value of using a low-fidelity mock-up early within the design process. Materials such as foam core board can be easily cut with a craft knife and the advantage of using the hook-and-loop fasteners is that things can be easily changed and adjusted. The use of the mock-up allowed user feedback to be obtained early enough in the design process to enable the designers to make substantial revisions to their proposed design (Long and Hughes, 2011). These revisions led to a second more robust mock-up of the full cab that allowed the effectiveness of these changes to be confirmed (Long and Hughes, 2011). The teams involved in this project still use the foam core board desk as an example of how to use mock-ups throughout the design process.

Case Study 3 – New Train Design 2

In another train build project a train was going into an existing system with the crew changing between the existing fleet and the new train through a day. Thus, an important part of the review was to consider crew expectations and negative transfer issues. Users of this train included crew, passengers, cleaning staff, security and other ancillary staff. The suppliers conducted two mock-ups at different stages of the process.

First, the suppliers employed a three-dimensional CAD system as an initial review tool with users rather than lower fidelity physical representations. This type of model is presented as a three-dimensional graphical image using a CAD package. The fidelity of information presented can typically be manipulated by the designer and can involve partial or full removal of specific components or structures to allow better visibility of aspects of the model.

During the three-dimensional CAD review we observed that the users involved did not appear to be fully engaged in the process. The system was driven by a technician and the reviewers appeared to receive the information as a presentation of the proposed design rather than an opportunity to review and critique the design. Unfamiliarity with the technology used meant that it was difficult for the users to interpret the images into a real life scenario.

A second high-fidelity mock-up was constructed at a later date during detailed design. The second mock-up contained the majority of equipment that was to be present in the actual cab and was finished to the same standard as the final product. Where it was not possible to connect certain electronic systems to the display screens, screen shots were used or alternatively a mock-up interface was generated on the display (see Figure 4.4).

The supplier performed a detailed review with crew via a very structured administered questionnaire. Crew who participated were naïve; they had had no previous exposure to the design development. In addition, the client also reviewed procedures, particularly those involving new systems and technologies. Some changes were made to the design and others to the procedures. This showed the benefit of design and procedural review occurring in parallel. The mock-up was reviewed a second time by the same crew to ensure changes had not introduced any unforeseen/unwanted issues. For other users, a group review methodology was used. A group walked around the mock-up and then the group was brought together to report on good features and areas that may benefit from change. In addition to the review of the mock-up, user feedback was gained by the use of a small group of users' representatives throughout the design process to assist with design review.

The mock-up proved to be effective in allowing crew, passengers, cleaning staff, security and other users to review and comment on the ongoing train design. This review allowed a number of issues to be identified that needed to be addressed.

Figure 4.4   Example of a high-fidelity mock-up

Case Study 4 – Software Project Usability Review

A new software system was being acquired to facilitate the development and management of timetables. The business requirements for the system had already been identified and provided to the system developer.

At the time of the usability review the project was in the feasibility phase (equivalent to the detailed design stage discussed above), conducting a proof of concept (PoC). The focus of the PoC was to confirm whether the functional capabilities of the prototype were in line with the business requirements for the system. The objectives of the usability review were to:

- review the usability of the system in its current stage of design and development
- identify areas for improvement to ensure the final system design was optimised to meet business requirements, and to provide enhanced support to help users complete their primary task(s).

The prototype software was reviewed using a heuristic evaluation guide and two different software interface evaluation tools (a usability checklist and a desktop review). An overview of how each of these tools were used follows.

Heuristic Evaluation Guide

The heuristic evaluation guide (HEG) is an Excel-based tool that includes 291 criteria that are broken down into 10 usability heuristic design principles (Nielsen, 2005). These principles include but are not limited to visibility of the system status; the provision of the system to help users recognise, diagnose and recover from errors; whether the system matches the real world/users' world; flexibility for the user to make decisions; and the provision of useful help and guidance. For each of the criteria the assessor determined whether the criterion was either *Relevant* or *Not Relevant* to the system – taking into consideration the stage of the design process. Where the criterion was considered to be relevant, the assessor then made a determination regarding whether the system was *Compliant* with or *Not Compliant* with that criterion.

Following completion of the HEG the data were analysed to determine the level of compliance with best-practice design heuristics. The results were provided as a percentage against each of the 10 design principles. Five categories were used as a guide to the compliance ratings provided by the heuristic evaluation:

- *Very low*: Significant usability issues are associated with the interface.
- *Relatively low*: The system requires significant improvement in the areas identified as non-compliant.
- *Broadly acceptable/average*: The system is compliant with the majority of design principles, however, there is room for improvement to bring the system in line with best-practice.
- *Relatively high*: A good level of compliance with best-practice design principles; there are areas where the interface could be improved, however there are no significant issues identified within the interface design.
- *Very high*: There are limited improvements required to the system as no significant issues were identified during the review.

This is an assessor-based review that requires the assessor to be familiar with the system interface and core functions. For the most realistic output, a tool like HEG should be completed by a person who is a subject matter expert in the development and review of interface design. Use of this tool facilitates the identification of key areas of concern associated with prototype systems (e.g. lack of/insufficient online help).

Software Interface Evaluation Tools

Two software interface evaluation tools were used to support the usability review of the system, including a Usability Checklist and a Desktop Review. Both tools were developed by Bearman, McCusker and Hughes (2012) through the CRC for Rail Innovation. The application of these tools is discussed below.

Usability Checklist

The Usability Checklist requires the user to respond to a series of statements about the usability of the system. For each statement, the user was asked to indicate the extent to which they agree or disagree with that statement. Users were asked to provide their response on a five-point rating scale, where one (1) indicated that the user strongly disagreed with the statement and five (5) indicated that the user strongly agreed with the statement. In addition, for each statement, users were asked to provide a ranking of how important the feature was to them as a user.

The checklist consisted of a mixture of positive and negative statements. For positive statements, e.g. *Learning to use the system was easy*, a high score indicates a positive outcome. Conversely, for negative statements, e.g. *I found the system unnecessarily complex*, a high score indicates a negative outcome. Therefore, in order for the system to perform well, it would need to produce high scores on the positive statements, and low scores on negative statements.

A small sample of users took part in the user-based evaluation of the system. All participants had received a high level of training in the use of the prototype system and had already interacted with the system as part of the PoC. Each user was asked to consider their experience using the system and complete the Usability Checklist. It is important that users have sufficient training and experience with a system to ensure that they are able to make informed decisions when taking part in a usability review.

Desktop Review

The Desktop Review is an assessor-based tool that requires the assessor to observe users interacting with the system and, based on observations, respond to a number of statements about the usability of the system. As with the Usability Checklist, for each statement the assessor will provide an indication of the extent to which they agree or disagree with that statement using a five-point rating scale. Again, there is a mixture of positive and negative statements within the Desktop Review, for example, *Any alerts are presented visibly on the main page*, would require a high score to indicate a positive outcome. In line with the Usability Checklist, for each statement the assessors provided a ranking of how important they considered the feature to be.

Two assessors completed the Desktop Review for the system. Neither of the assessors had received any formal training in the use of the prototype system, however, they were both familiar with the interface and had some experience in navigating and interacting with it.

The evaluation of usability of the prototype system identified a number of areas for improvement within the proposed interface. By implementing the changes identified in the review, the organisation was able to ensure that the final system design would be optimised to not only meet business requirements, but also to support users to complete their primary task(s). The users involved in the review showed high levels of acceptance of the system, which can be attributed to two main factors. Firstly, they were consulted early in the feasibility stage to determine the system requirements – prior to the development of the prototype. Secondly, they were closely involved in the testing and usability review of the system, during which time their opinions and feedback were actively sought and taken into consideration as part of the ongoing design process.

Conducting an evaluation of usability at an early stage of the system development (i.e. during proof of concept) added significant value to the overall project. The cost of implementing changes early in the design process is far lower than implementing them at the end of the project, when the final system is delivered. In addition, user-focused tools facilitate early buy-in to the system from end users, thereby increasing the likelihood of overall acceptance.

Conclusions

Mock-ups can be an extremely effective way of gaining valuable user feedback and can be used cost-effectively throughout projects to improve design. Provided that the user feedback process is managed appropriately the information gained is invaluable for the success of the design and user acceptance of the final product. This can be of significant benefit to both the organisation and the designer. To enable the benefits from user feedback to be gained, the user feedback process must be structured, timely and continual – starting at the concept phase and running through to the detailed design stage. This process should be reflected in contractual documentation.

Mock-ups have advantages over drawings and computer models as users find them tangible, realistic and understandable, and as such they are a good mechanism for facilitating a joint understanding of the design and improving ownership by the users. The following provides a list of practical tips for enhancing the effectiveness of the use of mock-ups within the design process:

- Identify opportunities for the use of low-fidelity mock-ups early in the design process and only develop them to the required level of fidelity.
- Identify the key interfaces within the system and the data that is required, and formalise the data collection and analysis process for each mock-up.
- Ensure that an appropriate sample of users is involved in the exercise.
- Ensure that the mock-up is located in a convenient location for travel and is easily accessible.
- Ensure the use of mock-ups is tied into the design process and to contractual milestones.

References

Baxter M., and Hughes G. (2008). Engineering the RailCorp PPP train: safe, clean and reliable. *Conference on Railway Engineering, 7–10 September 2008, Perth, Australia.*

Bearman, C., McCusker, L., and Hughes, G. (2012). *Two Useability Tools to Facilitate Software Interface Evaluation.* Manuscript in Preparation.

Buxton B. (2007). Sketching User Experiences: Getting the Design Right and the Right Design. San Francisco: Morgan Kaufmann Publishers.

ISO (2004). ISO 6385 Ergonomic Principles in the Design of Work Systems (2nd edn). Geneva: ISO.

Krug S. (2006). Don't Make Me Think: A Common Sense Approach to Web Usability (2nd edn.). Berkeley: New Riders Publishing.

Long A., and Hughes G. (2011). Users and design review – what mock-ups offer. HFESA 47th Annual Conference 2011. Ergonomics Australia – HFESA2011 Conference Edition, 11:14. 7–9 November 2011, Sydney, Australia.

Nielsen J. (2005). *How to conduct a heuristic evaluation.* Retrieved from http://www.useit.com/papers/heuristic/heuristic_evaluation.html on 5 July 2012.

Rail Safety and Standards Board (2008). *Understanding Human Factors: A Guide for the Railway Industry.* London: Rail Safety and Standards Board.

Schumann J., Strotthotte T., and Laser S. (1996). Assessing the effect of non-photorealistic rendered images in CAD. Proceedings of Conference on Human Factors in Computing Systems: CHI '96, 13–18 April 1996, Vancouver, British Columbia, Canada (pp. 35–41). New York: ACM Press.

Snyder, C. (2003). Paper Prototyping: the Fast and Easy Way to Design and Refine User Interfaces. San Francisco: Morgan Kaufmann Publishers.

Standards Australia (2006). AS 4024.1901 Safety of Machinery Part 1901: Displays, Controls, Actuators and Signals – Ergonomic Requirements for the Design of Displays and Control Actuators – General Principles for Human Interactions with Displays and Control Actuators. Sydney: Standards Australia.

Wilson J.R., and Haines H.M. (1997). Participatory ergonomics. In G. Salvendy (ed.), *Handbook of Human Factors and Ergonomics (2nd edn)* (pp. 490–513). New York: John Wiley & Sons, Inc.

Wright K. (2007). Cardboard to computers: an evolution of design visualisation. In J. Wilson, B. Norris, T. Clarke and A. Mills (eds.), *People and Rail Systems: Human Factors at the Heart of the Railway.* Aldershot: Ashgate Publishing.

Chapter 5

Qualitative Research Rules: Using Qualitative and Ethnographic Methods to Access the Human Dimensions of Technology

Kirrilly Thompson
Central Queensland University, Appleton Institute, Adelaide, Australia

Introduction

This chapter has been written strategically for those who are interested in, or involved with, the application of new technologies in the rail industry. It is intended as an introductory guide for those who are largely unfamiliar with qualitative research methods. Qualitative research is particularly suited to understanding human behaviour within the social, cultural and local context. Within this context, behaviours, attitudes, beliefs, etc. acquire particular cultural meaning which is reinforced and negotiated through social relations with other humans and non-humans (such as technology). Given that the success or failure of new technologies in the rail industry occurs within a cultural context and through social relations, qualitative research is particularly relevant and useful.

Despite the *double entendre* or double meaning of the title, I am not interested in developing an argument for qualitative methods over quantitative methods. Nor am I interested in prescribing a rigid or rule-bound formula for conducting qualitative research. Rather, I prefer to work from a flexible 'horses for courses' or 'what works best when' approach which is driven by aims and outcomes. To help people make informed choices about when qualitative research might be most appropriate, this chapter provides discussion and guidance on the topics of what qualitative research is, what its benefits are, when it should be used, how it can be used alone or in association with other methods and how it can be used to evaluate human factors issues of new technologies in the rail industry. From a practical perspective, I discuss how to conduct qualitative research, how to collect data, how to analyse it and how it can be presented.

As a cultural anthropologist with experience in successfully applying ethnographic research methods to address issues in the rail industry (Mueller, Thompson and Hirsch, 2012; Rainbird, Thompson and Dawson, 2010; Thompson, Hirsch, Mueller, Sharma-Brymer et al., 2012; Thompson, Hirsch and Rainbird,

2011; Thompson, Rainbird and Dawson, 2010), I include a specific discussion of ethnographic research methods. Ethnographic methods can be applied to all forms of research, especially where aims and objectives centre on determining how people understand their world, how they interact with other humans and non-humans and how behaviours, attitudes and beliefs are made meaningful, and shared with and communicated to others. This knowledge is central to evaluating new technologies where there is a need to understand users: what users do, what they think they do, why they (think they) do it, how that has changed in the past, what users need and desire for the future, what their immediate and long-term concerns are, how resistant to or accepting they are of change and technology in general and what is required to successfully support the introduction and acceptance of new technologies in their workplaces.

What Is Qualitative Research

The idea of qualitative research is based on a simplistic binary opposition between the terms 'quantity' and 'quality', and 'quantify' and 'qualify'. The term 'quantity' refers most obviously to numerical quantification. By default, qualitative research has come to be understood as everything else; that which cannot be easily reduced to numbers or which ceases to be significant or meaningful in numerical terms. The term 'quality' can refer to things which are not immediately expressed as numbers, such as the different views that workers have of new technology, why they do or do not see a need for new technology and what contributed to the formation of those opinions.

In practice, however, the distinction is less clear and the terms qualitative and quantitative can be more usefully considered as descriptors of different aspects of research. For example, a survey may reveal that 30 per cent of workers in an organisation see no need for new technology whilst 60 per cent do see a need and 10 per cent do not care either way. These figures represent quantitative findings (percentages) to a quantitative question ('what percentage of workers see a need for new technology?') about a more qualitative phenomenon (introducing new technology).

However, these percentages fail to reveal *how* workers perceive technology in their practice and *why* they see or do not see a need for new technology. The numbers also fail to determine the malleability of such opinions or convey how respondents interpreted the question. Even simple questions are often enmeshed in complex workplace politics and how much knowledge (or even scepticism) someone has surrounding the intentions or applications of the survey or questionnaire. These are all factors influencing the interpretation of individual questions and the responses that result.

'Quantity' can answer questions of 'how many', whilst 'quality' can answer questions of 'why', 'in what ways', 'to what extent' and 'how'. The fact that the 'whys' and 'hows' can also be quantified is discussed below.

I began this section by using 'quantity' and 'quality' broadly as 'terms', before applying them to specifics such as 'findings' or 'questions'. This is because the terms 'qualitative' and 'quantitative' can both be applied, but not limited to any of the following typical stages of research in qualitative and quantitative frameworks:

1. Research questions
2. Research design
3. Approach, conceptualisation and methodology
4. Methods
5. Data
6. Analysis
7. Interpretation
8. Reporting
9. Review
10. Critique

Whilst the list is given above in a logical order, there is no need to conceive of two mutually exclusive pathways via which this design can proceed for either qualitative or quantitative research. That is, all stages do not need to be consistently and exclusively either qualitative or quantitative. Whilst it may be generally appropriate for a qualitative research question to be answered with a qualitative design, approach, method, data, analysis, interpretation and reporting (and *vice versa* for quantitative research questions), this is not necessarily the case. For example, units of qualitative data can be subject to counts and analysed and interpreted quantitatively (i.e. content analysis). Similarly, qualitative research questions can be addressed using the kind of experimental or positivistic research design commonly attributed to quantitative research, such as a pre-post intervention research design to identify the impact of new technologies on workplace practices (Thompson et al., 2010). In many ways, labelling research as 'qualitative' or 'quantitative' constructs a false opposition.

In fact, it is hard to imagine any strictly qualitative piece of research having no quantitative elements. At the very least, research is always conducted by a number of researchers, with a number of participants, from a number of groups, cultures or subcultures, across a number of sites, on a number of occasions and for a period of time. Likewise, qualitative dimensions are relevant to all forms of research. The significance of knowing how many workers support or are against the introduction of a new form of technology becomes particularly meaningful in qualitative terms; in relation to issues of significance to an organisation such as political climate, workplace inefficiencies, safety considerations, recent incidents and behaviour change initiatives. For instance, consider the hypothetical example of train drivers who were not consulted about any previous introductions of new technologies, or who voted against a technology which was introduced anyway, without their concerns being addressed or the introduction being adequately justified. These drivers may be resistant to the introduction of further technology, even if they are involved in consultation, because the legacy of their previous experience has led them to expect that their opinions will not be valued. In some cases, drivers may even form the opinion that research is a tokenistic attempt to make a process appear consultative when 'management have already made their decision' (see chapter 3, this volume). Without knowing the historical context surrounding drivers' experiences of change, or their perspectives on driver-management relations, their resistance could be misinterpreted (perhaps as Luddites). In this hypothetical, qualitative research can be used to understand the broader context for the interpretation of quantitative data about how many workers support or reject new technology.

The following points describe some salient characteristics of qualitative research in further detail. However, it should be noted that they are generalisations. They are neither definitive of qualitative research, nor exclusive to quantitative research.

- Qualitative research *may involve low sample sizes* that may not be statistically significant. Significance in qualitative research comes from the ability to identify in detail the values, attitudes, behaviours and beliefs that are significant in relation to the research aims or question. In answer to the question of how many people is enough, the response usually takes the form of 'it depends ...' (Baker and Edwards, 2012), or until saturation (described below) occurs. Guest, Bunce and Johnson (2006), for example, reported saturation within the first 12 interviews and the presence of metathemes within only six interviews. The smaller samples of qualitative research can enable a greater depth of understanding.
- Conducting qualitative research *can take time*. Building rapport with participants and establishing effective relationships that allow the researcher to identify and collect detailed information about the broader relevant context can take time that is difficult to estimate. Thoughtful practice based on effective and sensitive communication with thorough planning can prevent unnecessary delays. Whilst time may be taken to undertake qualitative research, its added ecological validity (the fact that data is collected in the real world context to which findings relate) can improve the long-term success of interventions, changes, practices, etc. Analysing qualitative research can also be time-consuming.
- *Data collection and analysis often occur simultaneously* in qualitative research practice. This is due to an iterative process of recording and reflecting being central to the interpretation and ongoing evaluation and refinement of research. As a result, research design or tools may develop throughout a project. In fact, change is often a positive sign that the researcher has been critical about their practice.
- Qualitative research can involve, but *does not require, a hypothesis*. Hypotheses are typically based on prior assumptions about what might be found which can limit what can be found. Qualitative research is

particularly suited to the exploratory approach of finding out 'what else might be going on'. This is especially useful when an intervention has not worked. In place of a hypothesis, qualitative research may commence with a research question, problem or basic topic of enquiry.

- Qualitative research is *often conducted in situ*. That is, research is conducted with participants in the setting under consideration, rather than in an 'artificial' experimental or laboratory setting removed from the context in which the topic of analysis (behaviour, etc.) usually exists (see 'ecological validity' above). The rationale for this is that whilst a qualitative researcher might be interested in a single behaviour or attitude, they typically desire an understanding of the broader context within which behaviours, etc. exist, alter, are impacted by and impact on. Conducting in situ research provides the researcher with an opportunity to see behaviours, practices, attitudes and so forth in their usual context. The presence of the researcher may impact the research scene, however, this may be an important trade-off for in situ experience. Moreover, the researcher's impact should be lessened over time as they build rapport and familiarity with participants.
- Qualitative research *values subjective facts*. This is discussed in more detail below.

This list of characteristics about qualitative research may challenge the beliefs of those who are more familiar with experimental research practice (e.g. statistically significant survey samples). If this is the case, it is hoped that they challenge the reader to consider other ways of conducting research and provide different ways of doing so. If the reader is critical of this list, or indeed of any other aspects of qualitative or ethnographic research, they would do well to equally apply the same concerns or critiques to more experimental or quantitative forms of research. For instance, all forms of data can be modified, selected and presented in ways under the control of the researcher. Moreover, all researchers are always involved in the interpretation of data. Facts rarely speak for themselves but are made sense of in ways that exclude other ways of sense-making. Being equally critical of all forms of research is important in offsetting the unnecessary and unhelpful presentation of qualitative and quantitative research as oppositional. Jordan and Dalal (2006) provide further response to the concerns that managers often have around engaging qualitative researchers.

Although qualitative and quantitative approaches are frequently perceived as being in opposition, they can be complementary and are often used in combination to provide 'triangulation', which is '… the combination of at least two or more theoretical perspectives, methodological approaches, data sources, investigators, or data analysis methods' (Thurmond, 2001, p. 253). By providing different perspectives on one research topic or approaching the same topic through different means, triangulation achieved by combining qualitative and quantitative methods can provide comprehensive and nuanced understandings. Combinations of qualitative and quantitative methods result in what is often described as 'mixed methods' research. However, qualitative and quantitative methods are associated with different approaches to 'reality' and 'truth'; qualitative with an acknowledgment that there can be multiple versions of reality functioning at a socio-cultural level (social constructionism), and quantitative with an assumption that there is a singular reality operating beyond that level but open to scientific documentation (realism). There are therefore different opinions on the ways and extent to which qualitative and quantitative methods can be effectively combined or mixed (Sale, Lohfeld and Brazil, 2002).

In summary, whilst talking about qualitative and quantitative research at an abstract level is a useful and often necessary heuristic for talking about different forms and elements of research, the reader is advised to be mindful that one form of research does not necessarily exclude elements of the other, particularly in relation to data and analysis. This is an important guideline for reading this chapter and considering its relevance to the reader's own practice and research. That is, a reader who considers qualitative research peripheral to their own skills, interests and intentions which they may identify with a quantitative approach, is advised to revisit their beliefs and consider potential points of intersection, overlap or complementarity. It is further advisable to consider qualitative and quantitative research as mutually inclusive. In the remainder of this chapter, I necessarily use the terms 'qualitative' and 'quantitative'. I add caveats where relevant. For the sake of not being repetitive, I urge the reader to receive those terms with an open mind, in light of the above discussion of the broadness and inclusivity of those terms.

When Would You Use Qualitative Research – What Is It Good For

Numerical answers associated with quantitative approaches are sometimes understood as reductions of complex phenomena. In contrast, qualitative research seeks to preserve the richness and complexity of daily life in its interpretations and presentation. This difference between reduction and richness is sometimes evoked in the description of qualitative or social science research as 'warm and fuzzy'. This phrase is based on two related assumptions: first, that objective, hard, cold, facts isolated from other variables and subjective human emotions are the most valuable form of knowledge; and second, that warm and fuzzy is by default less valuable, less scientific or worse – 'the lowest form' of research (Savage, 2006, p. 383). However, the value of different forms of knowledge can only be judged in relation to the aims of eliciting that knowledge, or its significance and implications (Becker, 2007). Where there is a need to know 'more than' or understand the context surrounding the answers to questions such as 'how many' or 'how often', qualitative research is ideal. If there are indeed 'warm and fuzzy' forms of research, then surely there are 'warm and fuzzy' questions to be asked? If those questions are worth asking, their answers are no less significant than answers to other kinds of questions – even if they are not meaningfully expressed in numbers.

As already noted, qualitative research is particularly useful for understanding phenomena in context and providing a depth of understanding (to beliefs, opinions, behaviours, etc.). It can be used to:

1. understand (*inter*)*relationships* between people and phenomena
2. identify and understand *complexity*, contradiction, diversity
3. identify competing and impacting *influences* in the field under 'natural' conditions
4. identify *patterns*, processes and flows
5. identify variety, dynamism and *heterogeneity* across space and time
6. gain an insider or 'emic' view (discussed in more detail below)
7. gain an *in situ* understanding
8. provide *depth* of understanding
9. understand the *relation* between what people do and what people say they do (revealed and stated behaviour)
10. *engage* end users, solicit feedback, endorsement, approval and 'buy in'.

Qualitative research can also be used alongside quantitative research to:

1. *triangulate* with quantitative research (discussed above)
2. *develop a survey or questionnaire tool*
3. *develop or refine a research question or hypothesis* that may drive an experimental or quantitative pathway
4. *interpret or contextualise* quantitative findings (that may have arisen from any form of research).

To achieve these ends, a two-phase qualitative-quantitative or three-phase qualitative-quantitative-qualitative research design can be adopted, as illustrated in Figure 5.1. Take, for example, the need to make sure that a new technology reduces the workload of train controllers. Qualitative research could be used to *identify* all the different types of pressures that train controllers face. These pressures could then be used to *develop* a more sensitive and relevant survey seeking to rank the most significant or frequent pressures. This was the approach taken in research into passenger crowding whereby findings from qualitative research (Thompson, Hirsch et al., 2011) informed the design of an online stated preference choice experiment (Mueller et al., 2012; Thompson, Hirsch, Mueller and Rainbird, 2012). Rather than undertaking a stated choice design based on the usual suspects of time, money and standing/seated, the earlier qualitative stage directed us to include different locations on the carriage as well as safety concerns, thereby providing a survey tool with enhanced ecological validity (Thompson, Hirsch, Mueller, Sharma-Brymer et al., 2012). Whilst qualitative findings were used to inform the quantitative survey, they were also significant in their own terms and served as an end in and of themselves (Hirsch and Thompson, 2011a; Hirsch and Thompson, 2011b; Hirsch and Thompson, 2011c; Thompson and Hirsch, 2011).

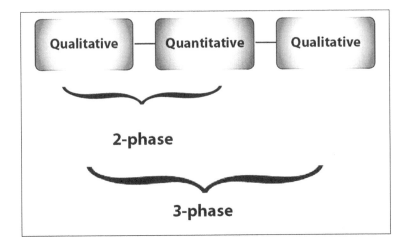

Figure 5.1   Two- and three-phase research designs involving qualitative and quantitative methods

To add a third phase of qualitative clarification, the findings or outcomes of the survey could have been discussed, user-tested or workshopped with the same workers during interviews or focus groups. This would *engage* the participants and target group, provide a greater understanding and interpretation of what the findings mean (for workers and for managers) and could be an opportunity to gain early *feedback* on recommendations, potential outcomes or changes to practice, as well as *endorsement* of those actions.

Finally, qualitative research can be used to identify target groups and to tailor strategies or initiatives appropriately. To continue the example above, a survey of the pressures that contribute to the workload of train controllers may identify that only some train controllers face all pressures during every shift. Where the available technological support may be limited, the survey could be used to prioritise those who should receive it first.

Do as I Say, Not as I Do

So far I have discussed the characteristics of qualitative research, and the relationship between qualitative and quantitative research. But what do people doing qualitative research actually do? Qualitative research methods are synonymous with interviews and focus groups where researchers talk to participants. However, observational methods also have a long history of use in qualitative research. One particularly good reason to conduct observational research can be found in the phrase 'do as I say, not as I do', which acknowledges the human tendency to act in ways contrary to knowledge about the ideal way to behave, i.e. to know the 'right way' to do something, but to all outward appearances, not do it. It is important to understand what people know, how they translate that knowledge into action, how they think they translate knowledge into action and how these may differ. For this reason, knowing what people think, feel, know and say provides only a partial account without also knowing what people do, and *vice versa*. Qualitative research is particularly suited to exploring and explaining the congruence and incongruence between what people say they do (i.e. their stated behaviour) and what people are seen to do (i.e. their revealed behaviour).

It should be noted, however, that cases of incongruence do not suggest that participants are contradicting themselves or being disingenuous. Often people are unaware of inconsistencies between what they say, what they do and what they believe they do or there is a form of internal logic through which there is no contradiction, such as a belief that someone has found an even better or safer means of achieving something. This is particularly important in research in organisations where people may believe they are doing the right thing or following procedure when evidence would suggest otherwise. In such cases, confronting workers without being aware of their perception of their own behaviour may invite defensive reactions or feelings of being misunderstood. Similarly, information-based behaviour change initiatives are unlikely to resonate with a target group who do not identify as needing to change (i.e. if you think you are doing something well, you are unlikely to see a need to read information on how to do it well). Knowing how people perceive their own actions before developing behaviour change initiatives may save time and money on ineffective campaigns.

Considering What People Do: Observational Research

For many reasons, it is important to know what people do. For example, anyone considering introducing a new technology may want to understand practices, interactions and issues with previous and existing technology. Likewise, anyone wanting to design a manual for using new technology may want to understand how (and if) manuals for existing technologies are used, developed, stored or consulted. Someone who is puzzled as to why a new technology is not being adopted or used in an expected way may wish to conduct observations to explain what is (and is not) happening 'on the ground', or to identify a barrier to best-practice that was not foreseen in the pre-implementation phases of introducing a new tool. In this instance, observation may reveal an impact on desired outcomes which would otherwise never have been anticipated or included in the design of other research tools.

There are also occasions when workers are unable to recall with sufficient detail their behaviour, especially when they have become so familiar with their tasks that knowledge is tacit and behaviour is taken for granted. In these situations, it makes sense to undertake observations of their everyday work processes *in vivo* (as they occur). In fact, one of the strengths of observation is that it can be conducted *in situ* (Wilson, Galliers and Fone, 2007).

The process of conducting field research typically involves 'getting in, getting on and getting out' (Buchanan, Boddy and McCalman, 1988). Getting into highly structured, hierarchical environments such as organisations and workplaces presents particular challenges. There are some environments (including workplaces) where an unfamiliar observer can be very obtrusive or disconcerting for workers and potential participants. Hospital bedsides are a prime example, as are instances where workers are used to carrying out their tasks with little to no company or overt observation, as is often the case with train drivers. In these instances, an observer may be more welcome if their presence is preceded by face-to-face interaction in the form of an interview or focus group, or by a previous research encounter such as a survey or questionnaire through which they can identify researchers. Other forms of introduction can include recruitment material with a photo of a researcher or a story about the research being placed in an appropriate newsletter or other media. Information sessions held for staff prior to the commencement of data collection can also be helpful in this regard (Thompson et al., 2010). In the same vein, there may be instances where a researcher will be advantaged by being introduced to a group by someone the group trusts (Palmer and Thompson, 2010). Where workplaces are characterised by worker distrust of management or an 'us versus them' mentality, a researcher would be advised to build a relationship with a worker who can then facilitate introductions. In these instances, it can be advantageous to select a co-researcher from the workplace to work alongside the researcher. This requires full support from management, especially with regard to the provision of time.

Observation in qualitative research practice usually differs from observation in other approaches like time and motion studies which are generally used to provide aggregated, numerical results. Qualitative observations tend to be less structured and include more detail of how something is done rather than what is being done and for how long (unless the process itself is the topic of enquiry). However, the two approaches can be successfully combined (Thompson et al., 2010, appendices 1–8). Figure 5.2 below provides a working example of a simple observational template that I have used in projects in various public and organisational settings. The table is one means of formalising fieldnote taking.

At the beginning of each observation, I record important details to characterise each session and provide context for notes. Where it is relevant, I sometimes also record the weather or other details important to a specific site or task. Extremely hot or wet weather, for example, may prompt train drivers to discuss particular concerns, such as needing to consume more water but finding it difficult to take toilet breaks, or having to manage the potential for slides due to wet tracks.

	Time	What is happening?	Significance?	Code/Theme/Label
1.	0650	I arrived at 06:50 and was greeted by the site manager who was not expecting me. I explained that I had arrived to conduct observations with drivers and she took me through two security enabled doors to introduce me to a driver coordinator.	Lack of internal communication?	Communication
2.	0700	The driver coordinator (DC) was expecting me. He asked me where I would like to start. I said that I was interested in watching the training session that was scheduled for 07:15. He said that he would take me to the driver simulation room.	Are there other forms of training to supplement the simulator?	Communication Simulator
3.	07:05	We went through a series of passages on our way to the simulator room. The D.C. explained that he had not yet had a chance to tell the drivers that I am coming to watch training.	No information session held with staff prior to commencement	Communication
4.	0710	The DC took me into a room of 8 drivers. 2 were female. They were sitting in chairs facing a whiteboard. The DC said that I was there to watch them train. I politely added a note about the purpose of the research and reminded drivers that participation was voluntary, that I was not there to judge how well they were performing in the simulator, but that I would like to know how they learnt to drive. I also told them that I would like to hear their thoughts about the training process.	Need to find out what information drivers were given before using the sim or how long they had been in training for. Need to remind drivers about voluntary participation when the DC is not in the room.	Training Gender
5.	0720	The DC's phone rang. He told me that he had to leave and directed me to take a seat at the front of the room.		

6.		One of the drivers (female) said that she helped with some research last year but nothing came from it.	Need to talk to management about circulation of findings.	Expectation management Communication Perception of research
7.		Another driver (male) said "yeah, I told them I didn't want to train in a simulator, I joined this place so that I could drive real trains"	Idea of 'real trains' may need to be challenged to get buy in from some drivers. Need to check how many drivers share this view in the organisation.	Simulation fidelity Expectation management 'Real' v Simulated train experience Anti-simulator
8.		I asked the group if they had the same feeling. They all nodded. I then asked 'is anyone excited about driving the simulator', and one person (male) called out "actually, I can't wait"	Variety of opinions re: simulator Challenge: Catering for various levels of interest	Pro-simulator Engagement

Standout observations

Many drivers look over 40 years of age. May impact their uptake of technology. Few women. Need to consider if there are gender-differences around using technology and the potential impact of learning simulator tech from a male/female or learning with a male/female colleague. Suggest one-on-one interviews to find out from the women themselves. If I was not female, I would recruit a female researcher to do these interviews so women felt more comfortable.

Personal reflections

As I was not expected by drivers, I will need to spend additional time with each one to ensure they feel relaxed with my presence when I watch them use the simulator. Will see if I can join this group during their lunch break to connect on a social level.

Further questions

Watch driver in note 7 use the simulator. Given that he is disinterested, it may be useful to see how this translates to his engagement. Does his attitude change after seeing and using the technology? What might this mean for managing expectations of simulators in the future and building interest prior to the use of the technology? How much influence might he have over other drivers' behaviours? Compare his experience with drivers in note 8.

Figure 5.2   Observational fieldnote template with example

The first column in the fieldnote template is used to number the note for easy cross-reference with the overall comments at the bottom of the page. The second column is used to record the time at which something occurs. There are usually no prescriptions or a need to record an activity at regular intervals. In qualitative research, units of time and duration of tasks are not usually analysed *per se* but rather provide a sense of duration, what time of day something occurs and the flow of time through one observational session.

I use the third column to record what is happening in my surroundings. Due to the diary or running narrative tone of writing notes like this, there is a tendency to write in present tense. However, this gives a vacuous, generalised feel that this is how things always happen. I am in favour of writing in the past tense, at least in the final presentation of findings, as a reminder that research is always time-specific and that practices and people are dynamic and ever-changing.

Observational notes are valid scientific data, as are the outcomes of all other qualitative data collection techniques (e.g. interviews, focus group discussions). To make sure that they are high quality, rigorous and systematic, it is important that they are written with as much detail or relevance as possible. Given that relevance can change in hindsight, more detail is better than less.

The aim of writing as much as possible in observational or field notes usually produces anxiety over 'where do I start?', 'what is most important to record?' and 'what do I observe?' The three strategies presented below may be of assistance, and can be easily modified in relation to the researcher's aims and objectives.

One rule of thumb for thinking about what to write about is provided by Bernard's (1995) identification of five major kinds of social science variables (pp. 113–14):

1. Internal states (e.g. attitudes, beliefs, values, perceptions, cognition)
2. External states (e.g. age, wealth, health, height, weight, gender, etc.)
3. Behaviour (e.g. reported/stated and observed/revealed)
4. Artefacts (e.g. Bernard refers to human residue such as waste and trash, but I extend this variable to include all the material objects with which people interact, such as technology and information such as user manuals and safety protocols)
5. Environment (physical and social environment, the weather, the social atmosphere, etc.)

By combining these five variables, Bernard identifies 15 different kinds of research projects and questions. For example, the question 'how does age impact worker adhesion to a machinery safety protocol?' combines an external variable with an artefact.

Spradley (1980, pp. 82–3) conducts a similar but expanded exercise by including the following nine variables in a Descriptive Question Matrix, reproduced in part in Figure 5.3 below:

Space	Space	Object	Act	Activity	Event	Time	Actor	Goal	Feeling
Object									
Act									
Activity									
Event									
Time									
Actor		*							
Goal									
Feeling									

Figure 5.3   **Spradley's Descriptive Question Matrix (adapted from the complete matrix by Spradley, 1980, pp. 82–3)**

Spradley drafts 81 example questions for the intersection of each of these variables according to the 10x9 matrix summarised above. The question in the box with a star would read 'What are all the ways actors use objects?' (Spradley, 1980, p. 82). To keep things simple, the reader may like to consider only those questions where the same variables intersect with one another in the matrix (see the shaded cells in Figure 5.3). For these 'like by like' intersections, the question simply becomes: 'Can you describe in detail all the places/objects/acts/activities/events/time periods/actors/goals and feelings?' When a research topic has been clearly defined, or when time is restricted, Spradley's Descriptive Question Matrix can be used to limit the number of potential questions from 81 according to their relevance, or at least accord more or less attention to particular variables. If a research topic is undefined, completing the matrix is a useful means of determining potential research questions in order to decide on their relevance and selection.

Another relevant heuristic or 'rule of thumb' covering similar dimensions in different terms is provided by the 'SHEL' model of human factors developed by Edwards (1972) in relation to aviation performance and safety. The model has been extended to 'SCHELL' with 'C' (Hofstede, 1983) for 'culture' (usually presented as an autonomous element rather than an unavoidable context for the entire model) and a duplicate 'L' (Hawkins, 1987) to convey social interaction between 'liveware' or people. Each letter of the extended SCHELL acronym stands for the following:

1. *Software*, such as information, policies, guidelines etc.
2. *Culture*, such as ethnicity (for example, Kouabenan, 2009; Stiles and Grieshop, 1999) but including local workplace cultures that normalise particular attitudes, values, beliefs and behaviours
3. *Hardware*, such as tools, technology, equipment and so on
4. *Environment*, such as climate, temperature, socio-political milieu, etc.
5. *Liveware*, referring to people, their attitudes, beliefs. Hawkins (1987) added another 'L' to refer to the interactions between people and the other elements of the model.

Bernard's (1995) Five Variables, Spradley's (1980) Descriptive Question Matrix and the SCHELL (Hawkins, 1987) model provide useful mnemonics when faced with the common anxiety experienced by researchers in the field of what to write notes about.

There are times where I immediately see the importance of what I am observing and describing in the third column of the Fieldnote Template above. In these instances, I make a comment on importance in the fourth significance column. I often consider significance broadly and include questions to guide my research or a note that I need further clarification. However, on more focused projects, notes of significance may be guided by succinct project aims and outcomes. As noted above in relation to relevance, significance can change in hindsight. By identifying more rather than fewer areas of significance, the researcher can often gain a deeper contextual understanding of their topic of study. I do not always complete the significance box for every observation. The significance column is effectively about analysing the notes or data in the third column. A fifth column can be included, as I have done above, for 'code'. Further detail of coding is provided below. When there seems to be a logical break in topic, I begin a new row.

The Fieldnote Template can be used *in situ*, or it can be written up later. Some researchers do not write notes at the time of conducting research, preferring to write them after data collection so as to be less of a distraction to participants, less obtrusive and to be more immersed in the event. My preference is to take notes *in vivo* and *in situ*, to ensure the transparency of my research role. Details that may compromise someone if they were to be read by another participant may be included in the *post hoc* write-up. I write up or expand my notes after an observation to complete and expand upon details that I was unable to record at the time and with an acceptance that memory fades quickly and that observational sessions can last longer than expected, in which case memory becomes less reliable. In situations where it is inappropriate to make notes, I commit as much detail to memory as possible and write up notes after the observation, such as in the car park, in a waiting room or even in trips to the toilet throughout a long observational session. Technologies such as voice recorders, videos, cameras and livescribe pens may be of assistance in this regard, where appropriate and when used ethically (i.e. with the informed consent of participants). No matter when notes are written, they should be expanded on soon after being initially written (preferably as an electronic document for ease of retrieval). This gives researchers time to reflect on what they saw and pay additional attention to the significance column, forcing them to ask the crucial question of 'so what' in relation to the aims of the study. Answering 'so what' entails identifying the significance of findings.

At the end of each observational template, I reflect on standout observations as a means of summarising what I learnt from a session. I may also make personal reflections such as 'must remember to dress more casually on days when I am not interacting with management much, as drivers seemed to find it difficult to relate to me', or 'there is no room in the simulator for me and the drivers to comfortably be together – consider if video might be a better option'.

After conducting several observational sessions, the researcher starts to develop a feel for the common and uncommon concerns, themes or findings and may be able to note an association with different sites, ages, genders and so on. Identifying common and repetitive themes across a sample often constitutes what is known as 'saturation'. Researchers often use saturation as an indicator that they have conducted sufficient observations or collected sufficient data. That is, they know when it is time to stop collecting data when they cease to collect anything new. However, saturation may be impacted by the amount of time available and the ability of the researcher to identify new data. Has saturation occurred due to a lack of perceptive probing? Are you probing to find new data for the sake of it? With this in mind, saturation needs to be considered in relation to the aims of the project. That is, 'is any new data being identified that relates to the research question?' If the answer is no, the researcher should reflect on their chosen method, its execution and their personal research style. If the answer is yes, the researcher should consider collecting more data to enable them to differentiate the most common or dominant findings across different drivers/situations/technologies etc. as relevant to the research question. Overall, saturation – like research design – should be considered in relation to the specific aims of the research project.

The reader may have noticed from the above example of a fieldnote template that I was involved in more than mere observation of this hypothetical situation. I also actively engaged participants in discussion. Whilst observation is an ideal means for determining what goes on and provides the researcher with an opportunity to watch behaviour that more or less approximates the 'everyday', there can be great benefit in asking someone to explain what they are doing and why.

That is, although researchers do not want to overly prime or influence participants, neither should they ignore opportunities to simply ask people what it is they want to know. It is important to create the conditions necessary for data to emerge on a topic, with as little interference, probing or priming as possible. However, if data is not forthcoming, the opportunity should be taken to ask participants directly and without leading questions. In the presentation of findings, responses can then be framed with the caveat 'when asked directly about …' as opposed to 'without prompting, participants commented …'

From the Horse's Mouth: Considering What People Say

One of the justifications for using observational techniques is to avoid making assumptions about *what* people do and *how* they experience their worlds and workplaces, particularly when those impressions are based on what people say or think they do. At the same time, the most obvious critique of relying on observational methods is the risk of making assumptions about *why* people do what they do, *how* they understand and rationalise it, or even *if* they do at all. Regardless of how nonsensical, illogical or contradictory someone's behaviour *appears* there is most likely a form of internal logic at work whereby that behaviour makes perfectly good sense (at least in the time and place under observation). In order to change behaviour, it is necessary to understand how that behaviour makes sense for, and is meaningful to, the individual, and how it may be reinforced by or shared amongst a social or organisational group and how it is informed by the broader cultural context.

The risk of assuming motivation for observed behaviour is equally if not more salient when the observer is familiar with the situation being observed. This is why it can be particularly challenging to conduct good observational research within one's own organisation, especially if someone works or has worked in the same area or role as that being observed (e.g. a manager who is a former driver). There are certainly benefits in being familiar with the tasks, environments and people being observed. Such familiarity can enable the researcher to appreciate the nuanced and political dimensions of a workplace and draw from pre-existing networks. It can also facilitate access and entry to a research field (also known as '*entrée*'), assist with acceptance by participants and enable the building of rapport with participants. However, as cautioned above, these benefits can also disadvantage researchers by introducing assumptions, biasing the researchers' decisions and compromising critical reflection on observations.

As a result, when conducting research in one's own organisation or group, relying on observation alone is discouraged. For this reason, I recommend supplementing observation with interviews, focus groups or other data collection methods (with appropriate ethics procedures in place). This provides an opportunity for the researcher to clarify observations with participants (e.g. 'I've noticed that you … can you tell me more about that?') and allows the participant to comment on the researcher's observations. A researcher may also consider collaborating with other researchers and comparing observations collected from across a research team.

Of course, all researchers should be aware and critical of the assumptions and presuppositions that they bring to any research encounter – be it familiar or strange, or the strange that all too quickly becomes familiar. They should also be aware that trade-offs are inherent in all research decisions, such as expedited *entrée* at the risk of making assumptions. Amongst other things relevant to particular research projects, what counts is:

1. that the trade-off was the best one in relation to the research questions and aims
2. that other external forces are taken into account, such as time
3. that strategies are in place to mitigate trade-offs
4. that trade-offs are acknowledged
5. that trade-offs are considered when interpreting data.

To understand the human, social and cultural environment in which technology operates and behaviour occurs, it is important to know what people think, say and feel. This can be what they think, say and feel about many things, such as management, their job, their value to an organisation, their opinions of technologies, their experience of change management processes and so on. The reader may respond to this assertion with some confusion and comment that what people think are just their subjective opinions, i.e. they are not statistically significant. However, regardless of whether subjective opinions are subjective or not, what people think matters. This is the premise of what is known as the Thomas' Theorem: 'If men define situations as real, they are real in their consequences' (Thomas and Thomas, 1928, p. 572). For example, if train controllers believe that they will be severely reprimanded for reporting their own fatigue, they will be unlikely to report fatigue, even if their belief is unjustified or unprecedented. Such subjective facts as thoughts, attitudes, opinions and beliefs can be identified, collected, analysed and reported in the same way as any other kinds of fact or unit of research data.

The best ways to elicit people's thoughts, feelings, opinions and experiences is simple – just ask! Asking can take many forms, including surveys and questionnaires where the researcher need not be physically present to do the asking. Questions can be open-ended or closed depending on the ability to answer questions with open responses or to select from pre-arranged responses (such as multiple choice answers or responses on a Likert scale). Asking can also take more conversational forms, through focus groups and interviews where the emphasis is on the researcher as listener (Gerard Forsey, 2010). These can be structured, semi-structured or unstructured and moderated with more or less guidance from the moderator or interviewer. Some focus groups are run without a moderator altogether, being participant-driven according to some brief instructions. Moreover, asking can be driven by a bottom-up, exploratory, inductive approach where the researcher attempts to create an interaction where the participant drives the discussion according to their concerns, or they can be more top-down and deductive, where the researcher seeks to solicit information on, and keep the discussion firmly centred on pre-ordained questions and topics that they expect to find. As with the terms 'qualitative' and 'quantitative', the terms 'inductive' and 'deductive' are not always mutually exclusive. Even the most inductive research approach where the researcher 'goes in' to find out what 'might be there' is usually driven by a pre-identified need or topic of interest which provides some scope to a project (Morse, 1999). The rationale for one approach or a combination of approaches depends on several factors, including:

- the extent to which the interviewer or focus group moderator has conceived of a 'script' for discussion. For example, if there is very little known about a topic or issue, it may be more useful to develop conversation around a topic rather than specify questions.
- the extent to which the topic or scope of discussion has been defined. For example, research around the impact of technology on workload could involve questions around time on task, time off task, rest and recovery opportunity betwixt or between shifts, workplace information and communication, workplace culture and so forth (factors that could characterise a deductive approach). However, there could be other unanticipated factors affecting workload which can be identified through an inductive approach (such as the stressors associated with particular stations, platforms and lines).
- the extent to which the broader context for a topic is desired and can be addressed if found to be significant. For example, an inductive or exploratory design to an interview around technology use may solicit information about workers' unhappiness with other or peripherally-related issues, such as workload or manager-worker relations. Researchers should consider the implications of collecting such information on participant expectations of the aims and outcomes of the research. Setting unrealistic expectations, even inadvertently, could contribute to workers' feelings of alienation or disenfranchisement and may negatively impact the reception of the research findings as well as jeopardise participation in further research.

Overall, the skills required for conversational research are those required for any effective social interaction (see, for example, Shank, 2002), but the researcher's rule of thumb is to talk only as much as is necessary to develop the conditions under which the participant feels free to do the majority of talking.

Interviews in situ

In some projects I have had insufficient time or permissions to conduct observations *in situ*. In those instances, and where practicable, I have conducted interviews *in situ*; that is, in the workplaces under consideration, usually at people's workstations. This provides the opportunity to ask someone to provide an assisted description or re-enactment of their typical workday, tasks or role (as relevant to the research aims and objectives). This functions as a kind of situated 'armchair walk-through' question that I typically include at the beginning of interviews. Another kind of 'armchair walk-through' is the one taken by a researcher developing a research design, and imagining how they would put that design into practice (where would they locate themselves, who would they talk to, what challenges might they face, etc.) (Morse, 1999). The 'armchair walk-through' question takes the form of 'can you please take me through a typical shift/train walk-around/safety check, etc?' Conducting an armchair walk-through *in situ* often results in the interviewee physically taking me into the spaces where they work, explaining their role, identifying the location of important or essential equipment, information or technology and demonstrating how it is used. Walk-through questions are not only important for gathering data about tasks and providing physical cues for prompts around questions such as 'how often would you do that?' or 'when was the last time you had to … [use the emergency button]?' They also serve as excellent warm-up questions at the start of a research interaction (in observations or interviews) that help to build a relationship between researcher and participant (*rapport*). Even more importantly, the situated walk-through diffuses the power dynamics inherent in two people seated at a table where one does the asking, one does the responding and the one responding has to articulate in words that which they often do in tacit and embodied ways that they have not thought about – let alone put into words – before. By providing the interviewee with the environmental cues and props to supplement their verbal interactions with the researcher, the tone can be set for the participant as expert and the researcher as apprentice, thus helping the participant to feel confident in what might otherwise be an intimidating situation. The role of participant as expert is discussed in more detail below, in relation to ethnography.

There are also instances where it is inappropriate or unsafe for a researcher to carry out any observations or walk-throughs *in situ*. There are methods to collect observational data which do not require the physical presence of a researcher, including audio-visual recordings (with appropriate ethical approvals). In some projects, observation may simply not be an option, especially in highly sensitive contexts subject to extremes of confidentiality or security, in which case other non-observational methods can be used (interviews, anonymous surveys and so on). A verbal walk-through *ex situ* can be helpful in these instances.

Mixing the Observational with the Conversational

Combining observational methods with conversational or – more accurately – listening methods constitutes 'triangulation' of research methods and produces a more in-depth and rich account of what is going on in a given context (see also Wilson et al., 2007). As noted above, a combination of methods can produce a rich and comprehensive account of what people do, what they say they do, why they (think they) do it and how they (say they) feel about it (amongst other things).

The order in which observation and 'conversation' occurs can depend on many factors, including practical considerations. In general, I find it useful to conduct observation first (or the situated walk-through discussed above). Based on observations, I am able to design questions in surveys/questionnaires/focus groups and interviews to clarify, contextualise or otherwise make sense of what has been observed. Moreover, observations provide a small amount of familiarity with work practices and conditions that enable me to build rapport and trust with research participants. For instance, by starting interviews with statements such as 'I've been watching workers in here for nine days now and I've noticed that you have to shout to hear one another over the machinery', I am able to demonstrate my interest in what it is like to work in the participant's environment. This was the approach taken in the *Crowding* project where focus groups with rail passengers were undertaken after the researchers had travelled in passenger trains for one week beforehand. For example, we observed that passengers congregated around the vestibule rather than spreading out along aisles or taking seats. Focus group participants clarified the multiple reasons for this behaviour, including needing to get fresh air from the doors opening at each stop, having luggage that could not easily be moved elsewhere and being in school uniform and feeling that adults would stare with the expectation that they should be given a seat (particularly the case in one state where those travelling on school concessions were required to relinquish their seats to full-paying passengers) (Thompson, Hirsch, Mueller, Sharma-Brymer et al., 2012).

In another project exploring driver perceptions of workload and fatigue, focus groups were held at the same time that researchers were conducting in-cab observations and interviews with drivers. The rationale was pragmatic; there were limited opportunities in a two-week fieldwork period to hold 90-minute focus groups with six to eight drivers who had volunteered to participate.

Participant Observation and Ethnography

One form of observation where the researcher can be particularly engaged with the objects and subjects of research is known as participant-observation. Participant-observation is considered the hallmark of socio-cultural anthropology, although its application in other fields is increasingly popular, including marketing (e.g. Keaveney, 2008), cultural studies (e.g. Zimmermann, 2007) and design (e.g. Ball and Ormerod, 2000; Clarke, 2011). Participant-observation is one of the multiple tools under the umbrella term 'ethnography', which also includes censuses, questionnaires, surveys, genealogies, interviews and focus groups. However, participant-observation has become synonymous with ethnography. 'Ethnography' literally means 'people-writing', deriving from the Greek terms for 'people' (*ethnos*) and 'to write' (*grapho*).

Ethnographic approaches are largely based on an understanding that meanings differ across cultures, contexts and situations and that it is inappropriate to consider behaviours, attitudes, beliefs and so on in isolation from their cultural context. Using external or 'outsider' perspectives to interpret, evaluate or judge the behaviour, views etc., of someone from a different group constitutes an error of ethnocentrism. To avoid being ethnocentric, the aim of ethnography and socio-cultural anthropological practice is to gain an insider's understanding of what other people do and what meaning they attribute to their actions. This insider view is often referred to as an 'emic' perspective and understanding as opposed to an 'etic' or outsider perspective and understanding. An example could be where managers and human resource officers believe that train drivers are not consuming recommended quantities of water because they have ignored health and safety recommendations around remaining alert. This would constitute an 'etic' interpretation of driver behaviour. An 'emic' interpretation from drivers themselves could be that fluid intake is deliberately restricted due to insufficient opportunities or facilities for toileting.

In short, the premise of doing ethnography is 'to walk a mile in someone else's shoes' but to the extent that someone could explain the internal logic and experience of that walk to someone who has never done it, or who may have no knowledge of what a walk of any kind entails. The extent to which this can be fully accomplished is as questionable as the extent to which any one person shares the same world view as someone from their own culture. What is important is the acknowledgment that there are other, different, ways of seeing, which are no better or worse than one's own, and that an attempt is made to understand different perspectives without judgment. Of course, this ideal is challenged in workplace contexts where there is a clear need to judge practices as safe or unsafe.

Emic understandings can be gained through formal and informal interviews, focus groups (usually based on open-ended questions), other face-to-face interactions and open-ended responses to survey questions. They are often easier to identify or elicit through *in situ* research such as participant-observation. Participant-observation involves, quite literally, participating and observing. It requires a level of rigour and detail in the act of recording and writing up that is belied by the apparent informality of its practice. This requires a tricky mix of observational distance coupled with intimate participation (Sands, 1999, p. 25). Participation is required to be able to understand what it is like to be someone else, such as a train driver, manager or controller. However, some observational distance is required to be able to consider the experience analytically and objectively.

Spradley (1980, pp. 58–62) identifies five different degrees of participant-observation from observer-only through four different types of participation:

1. Observer (not necessarily present or participating)
2. Passive (bystander, spectator, loiterer)
3. Moderate (participates, but low competence)
4. Active (participates to minimum competence, self-study)
5. Complete (pre-existing participant, competent in the situation being studied).

These different degrees of participant-observation can be easily imagined along a spectrum from observer to complete participation. The risk of misinterpretation is greatest at the 'observer' end of the spectrum whilst there is a risk of uncritical taken-for-grantedness at the 'participation' end of the spectrum (see Figure 5.4 below).

The extent to which full participation can occur at any given time is dependent on many factors, including the participants, the researcher, the context, the environment, time (the researcher may become more or less accepted over time)

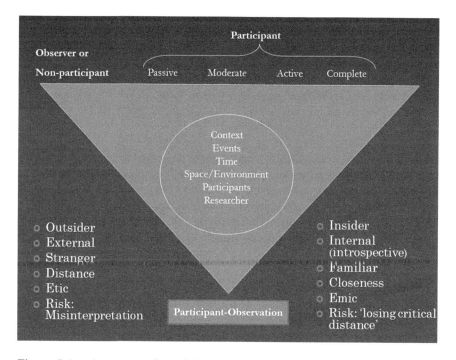

Figure 5.4   Spectrum of participant-observation showing characteristics at either end and incorporating Spradley's types of participation (1980, pp. 58–62)

and the milieu (recent events or other research projects). There are also practical and safety considerations impacting participant-observation. For example, it is unlikely that a researcher could fully participate in being a train driver, although they may be able to drive a train simulator or accompany a driver on part or whole driving shifts, as I have done in previous research with train drivers (Rainbird et al., 2010; Thompson, Rainbird et al., 2010). In my doctoral research on mounted bullfighting in Southern Spain, I rode a horse in a private bullring around a bull, but did not fully participate in bullfighting (Thompson, 2010a; Thompson, 2010b).

For traditionalists in socio-cultural anthropology, ethnography and participant-observation involve sustained participant-observation with a group of people for at least 12 months to capture seasonal variation in practices and to understand any single, academically reduced unit of study (e.g. a single behaviour, view, belief, action) within a broader and essentially irreducible network of mutually inclusive relations.

Traditionally, socio-cultural anthropologists produce a manuscript referred to as an 'ethnography' to convey the lived worlds of 'X' people in 'Y' location, often with a particular focus (see for example Agar, 1973; Malinowski, 1961; McHale, 2003). Thinking about the conceptual bracketing of ethnographic research in relation to specific people, places and phenomena is a pertinent reminder that all research is bound in time, place and space. Regardless of whether or not research findings are generalisable, they are all based on a specific group of people, in a particular place at a particular time and in relation to particular phenomena. In an era marked by globalisation, it is too easy to consider organisational environments such as the rail industry, managers, drivers, train controllers, maintenance, support and so on as the same, regardless of their location. Whilst they may share commonalities, there are undoubtedly local, regional and specific workplace cultures that provide points of difference.

There are clear challenges in applying a traditionalist sense of ethnography to studies of workplaces and organisations. Conducting sustained research for more than 12 months in an organisation is often impractical, cost-prohibitive and probably extemporaneous to the needs of highly focussed research questions. So what makes an ethnography an ethnography, or what makes something ethnographic? 'Walking a mile in someone else's shoes' is at the heart of ethnographic approaches and qualitative research in general (see Agar, 2006; Roth, 2006 for further debate). For example, managers who routinely accompany train drivers in their cabs several times a year are not conducting sustained, cultural immersion on a grand anthropological scale consistent with the views of traditionalists. However, as with the list of research stages presented above, their desire to build relationships with drivers and gain an insight into what it is like to be a driver in their organisation can be considered ethnographic.

Even those not engaged with face-to-face research need to be aware of the role of researchers in any piece of research. Presentations of the worlds of insects, the laws of nature or the workings of machines are inescapably human interpretations of those worlds.

What Makes Ethnographic Research Different

In the preceding section, I detailed what I believe is at the core of ethnographic approaches: the desire to elicit insider understandings. Whilst that which is produced is not necessarily an ethnography, at least in the traditional sense of the term, it can at least be understood more or less ethnographic. There are two particular stances that distinguish ethnographic and qualitative research which I feel are important to discuss in more detail. These relate to the role of participant as expert and the researcher as a research tool. However, it is not my intention to infer that these hallmarks are exclusive to or definitive of ethnographic research.

Participant as Expert

As noted above, ethnographic research prioritises the insider or 'emic' point or view. As such, the participant is recognised as the expert on their point of view and those of the social or cultural group with which they identify. This positions the researcher as naïve observer or professional stranger in a 'student-child-apprentice learning role' (Agar, 1980, p. 194). When I conducted ethnographic research into workload and fatigue amongst Australian metropolitan train drivers, I represented the Centre for Sleep Research. The Centre was selected to conduct this research because of its recognised status as expert in sleep and fatigue. However, in order to understand how drivers recognise, experience and respond to feelings of fatigue, I had to recognise that drivers were experts and I was their apprentice. I conducted participant-observation by talking to drivers and accompanying them in their driver cabs and lunch rooms so that I could learn from them and attempt to see the world through their eyes (Rainbird et al., 2010; Thompson, Rainbird et al., 2010). This 'emic' approach of the qualitative research stage contrasted the second quantitative stage of the research project which involved sleep researchers taking quantitative workload and fatigue data from drivers via activity monitors and making 'etic' judgements about whether or not drivers were fatigued. However, the two different research methods used in the overall project design ensured that factors identified as contributing to workload and fatigue in the qualitative phase were included in the evaluations made as part of the second stage.

Because the participant is the expert, there is no right or wrong. It is from this perspective that ethnography can put into action a commitment to be non-judgmental. However, that stance is not a given. It requires that the researcher identify and suspend their own value-laden judgments of which they may not always be aware. This can be particularly difficult for someone conducting research in their own organisation, where they may represent an organisation and feel obliged to share its beliefs about what is good, bad, right or wrong. One strategy around suspending judgments could be for the researcher to write down all of their opinions in the first column of a two-by-two table. In the second column, they can challenge themselves to see alternative points of view, or record instances where participants have expressed an alternative point of view, and justify why that view matters. Identifying specific judgments is arguably harder to do. However, given the emotional content of judgments and the emotional reaction to seeing behaviour and hearing accounts that conflict with our own judgments, it is possible to use one's own emotional reactions (disbelief, anger, frustration, disdain, etc.) as a trigger to reflect on and identify one's own judgments and beliefs.

At this point, the reader may be protesting 'but there is a right and wrong way of doing things – especially where lives of drivers and passengers are at stake'. However, holding this view makes it difficult to build rapport with participants and to gain insider points of view. It also makes it difficult to develop effective change management strategies that resonate with workers rather than inciting a defensive attitude or one that perpetuates the 'us versus them' mentality that characterises the operations-management relations of many organisations. It may be necessary to suspend judgments of right and wrong for at least the process of data collection, or to rephrase those judgments in questions such as 'in what ways are these behaviours justified by workers, seen as acceptable, learned, etc?'.

Managers play an important role in making sure that the selection and use of new technologies meets a variety of specifications, needs and budgets. Managers are the experts of these domains. However, even if a manager has been a driver or other user of new technology, it is the users of technology that are, or will become, the experts in how that technology is used every day by primary users. The acknowledgment of participants' roles as experts should not be tokenistic. Rather, it is a humble recognition that users are exceptionally expert.

Researcher as Tool

The second stance that characterises ethnographic and qualitative approaches is the one whereby the researcher acknowledges and critically reflects on their role as a research tool. This awareness is referenced in the idea of researcher reflexivity whereby researchers are encouraged to be reflexive about the impact of personal dimensions such as age, gender, nationality and religion at all stages of the research process. In particular, the gender of researchers needs to be taken into account in the research process (as discussed by Gurney, 1985, p. 44). There may be risks associated with being male or female in certain situations. Should you conduct research on your own or in a pair, for example? If you conduct research accompanied, should that be with a male or female researcher? As occurs in all social interactions and 'presentations of the self' (Goffman, 1959), 'speakers may be concerned to produce themselves as a certain type-of-person' (Rapley, 2001, p. 308). Is the gender of participant and researcher likely to result in exaggeration or downplaying of particular behaviours being researched? Moreover, the research product or ethnographic account is considered to be co-created by (and co-creative of) researcher and participant (Aull Davies, 1999; Coffey, 1999; Steier, 1991). For example, Rapley (2001) considers the interview to be a context-specific form of social interaction, not a representation of the topic of conversation. Rapley (2001) goes further to point out that the researcher is central to the production of 'talk' in qualitative data collection methods such as interviews. Such considerations of the researcher as tool should be anticipated and considered during the phase of research design.

When the Researcher is a Manager: Ethnography at Home

Above I noted the challenges to conducting good observational research within one's own organisation. The reasons for this related to the premises behind the formalisation of anthropology in the nineteenth century. Around this time, ethnographers and anthropologists were concerned with studying the cultural 'other', usually people considered 'native', 'uncivilised' and 'exotic'. Given that culture is, by definition, taken for granted and naturalised, it is often easiest to recognise from the outside, when one feels like the proverbial 'fish out of water'. Once recognised, the aim of the ethnographer is to be able to understand local culture in its own terms and be able to explain and communicate that cultural understanding to others. This was an advantage to most early ethnographic research which was conducted with 'other', non-Western cultures. The idea of doing ethnography far from home with a single group of people whose actions were fairly bounded has been challenged by two movements: multi-sited ethnography (Falzon and Hall, 2009) and ethnography 'at home' (Peirano, 1998) within the same culture with which the research identifies or is familiar. Those conducting research in organisations, corporations, institutions and workplaces may be conducting these two forms of ethnography. This brings its own unique challenges coalescing around the ability to develop a critical stance to identify the 'taken for granted'. Nonetheless, being 'at home' does not necessarily entail familiarity. For example, I conducted ethnographic research on the role of alcohol in the culture of Australian Rules football fans (Palmer and Thompson, 2010; Thompson, Palmer and Raven, 2006; Thompson, Palmer and Raven, 2011). The research was conducted in my own city of Adelaide in South Australia where I was very familiar with the culture but relatively unfamiliar with the group being studied: football fans. The inverse was true in my research on mounted bullfighting in Southern Spain where I was familiar with horse-riding and horse husbandry but the cultural context and the specific equestrian activity were strange (Thompson, 2004; Thompson, 2007a; Thompson, 2007b; Thompson, 2010a; Thompson, 2010b; Thompson, 2011). Whether doing ethnography at home or away, the ethnographer needs to engender, balance and reconcile familiarity with distance, or the familiar with the strange.

Depending on the pairing of the researcher's background with the research context, distance and familiarity will require more or less critical awareness. When ethnography is undertaken 'away', in a cultural environment foreign to the researcher, cultural differences are salient. The researcher's role then entails making the strange familiar or knowable and understandable to an audience with little familiarity. When ethnography is undertaken 'at home', cultural elements may be overlooked or taken for granted. In this case, the researcher must identify and challenge their ethnocentric assumptions by constructing the familiar as strange (see for example, Mannay, 2010; Miner, 1956). This was the basis of an ethnography of my own research group (Thompson, 2013).

When research is conducted by a manager rather than a consultant researcher, managers should be particularly conscious of the impact of their position in the organisation on the power dynamics of conducting observational research. They should make it clear, for example, that they are interested in knowing what it is like to be a worker (or if they have previously been a worker that they are interested in how practices have changed). They should also make clear what, if any, consequences may arise depending on the behaviours that may be observed. As the aim of conducting observation is to identify typical practices, managers conducting observations should ensure that workers feel comfortable to behave as per 'business as usual' even if they are aware that business as usual may not be business as prescribed/desired by policy/procedures/management. Depending on the sensitivity of the topic of research, managers need to be aware that workers are likely to reveal different kinds of information to an external researcher. If managers outsource research to address this concern, they should be aware that external researchers may be privy to unexpected opinions, behaviours etc. The ethical aspects of this should be considered and addressed by the consultant as well as the organisation and appropriate participant representatives. In fact, the more sensitive those opinions and behaviours, the more likely it could be that managers are surprised by the researcher's findings. Prior to commencing research, managers and researchers should have a lengthy discussion about how to deal with findings that are unexpected or 'negative'. As noted above, given that one of the benefits of qualitative research is the identification of unanticipated findings, managers and researchers are also advised to discuss original scope and potential for that to alter or be politically influenced as the research proceeds.

Analysing Qualitative and Ethnographic Data

After having collected some data, written notes in an observational template, made notes during interviews or focus groups, recorded those events or had transcripts made, the researcher is faced with the task of analysis. Much has been written about analysing qualitative and ethnographic data (Richards, 2005; Saldana, 2009). Green et al. (2007) offer a basic starting point for the thematic analysis of interview data. They identify four steps in the process:

1. Immersion in the data
2. Coding
3. Categorising
4. Generation of themes, where themes can be understood more generically as findings.

Contrary to many academic portrayals of codes and themes emerging from the data, analysis takes work on behalf of the researcher. So as not to leave the reader of this chapter stranded, I will outline the analytic process that I have used in my own research. In very simple terms, analysis can be understood as the act of breaking something down into parts/units of data, identifying the relationships between parts and then reconfiguring the parts into a narrative that addresses the research question or communicates a finding. In qualitative research, units of data might derive from observations, interviews, reflections, images, material objects and so on. Using the idea of generic units of data, stages of qualitative data analysis (QDA) can be conceived of as follows:

1. Review all the data collected.
2. Re-familiarise oneself with the data.
3. Identify units of data.
4. Name, label or code those units of data.
5. Categorise your data according to the code.
6. Develop an understanding of each code, including internal breadth and diversity.
7. Identify the relationships between codes, including patterns, processes or types.
8. Order your codes logically according to the identified relationships.
9. Articulate the relationships between codes. That is, describe the inter-relationships.
10. Consider those codes that are not accommodated by your description. Are they irrelevant or unaccounted for?
11. Develop a model or explanation [theory] that orders codes cohesively.
12. Critically reflect on the explanation that you have developed. Would participants agree or disagree, for example?
13. Recognise the significance of the theory (to your workplace or perhaps to the literature). That is, answer the implicit 'so what?' question.

The steps outlined above are a guide only. They will need to be modified or extended according to your own research. They will most likely be iterative, with movement back and forth. They can be carried out with specialist software such as the NVivo Qualitative Data Analysis Software (developed by QSR). However, they can also be achieved using spreadsheets (Meyer and Avery, 2009), tables in word processors, or just a notepad and pen or highlighter.

The fieldnote template can be adapted to support the formation and identification of codes. As described above in the description of the suggested Fieldnote Template, an additional fifth column can be added in which the researcher can input single words, descriptors or phrases that describe the major finding in each row. These words and phrases could then become codes/themes. A code is simply an abstract label that describes the thing or things that the data represent. It can help to imagine a code as the label you would write on a shoebox that you were going to literally store your observations or data in. It might be 'resistance (to technology)', or simply 'technology likes' and 'technology dislikes'. By reducing each observation to a number of units of data, and considering them as such, it becomes possible to label them with codes. If many codes are identified, they may be sorted according to relevance or aggregated again into higher-order categories. The role of data reporting then becomes to explain the category (which can be discussed as a finding, a theme or a code) and through a process of disaggregation, explain its detail, depth and complexity (as sub-codes or whatever terms make most sense in the research context and audience). Overall, the process of QDA is often one of dis-aggregation of units of data from one form and their re-aggregation into one relative to the research aims, question or problem. Depending on the researchers' aims and what they do with the codes they identify, they may not use all codes.

By writing a paragraph about each code to solidify what it means and why it is significant and identifying and describing the relationships between them, the researcher can quickly identify significant findings. In a report form, the researcher then need only introduce the theme/code/finding, provide some evidence or data from research to support it and then summarise the significance of that theme/code/ finding in relation to the research aims and questions. Where multiple findings are presented, there needs to be an overarching argument or narrative that organises them in a logical order and cohesive structure.

Presenting Qualitative Research

Clearly, qualitative research is suited to addressing qualitative questions such as who, where, how, in what ways, to what extent and especially 'why'. Of course, quantitative research can also address these questions, but usually by different means and to different ends. In a qualitative research framework, research questions are addressed in ways that frequently take broader social and cultural context into account. However, presenting contextualised findings can be a challenge for succinct communication and information delivery. Qualitative research is often concerned with acknowledging the depth and detail of small samples whereby differences and variation of opinion or practice within a group are sought and valued. This can be difficult to present in organisational settings where a culture of executive summaries, graphs, tables, flowcharts and briefings are the norm (Jolly, Radcliffe, Smith, Nycyk and Andersen, 2005/2006).

Although I have gone to great lengths in this chapter to note that qualitative research is concerned with detail and context, it does not have to result in a dense report (or tome of epic proportions) that is too long and impenetrable for any time-stricken manager or executive to read and too dense for anyone else to bother. What a qualitative research product does need to do is convey and acknowledge the complexity in simple terms, with sufficient evidence to demonstrate that interpretations are sound and that the findings are based on a rich body of data collected carefully, rigorously and always with the highest regard for the opinions, attitudes and beliefs of participants. Quotations from participants convey this very efficiently. Ideally, they should not identify the participant and they should be selected to represent not only the dominant view but the full variety of views, including conflicting views.

Qualitative findings can be presented in the form of, but are not limited to:

1. models (e.g. showing the situations where a new technology is more or less likely to be useful, used or accepted)
2. typologies or taxonomies (e.g. of different driver types according to their perspective on technology, or of the different types of uses of particular technologies. For example, different kinds of people in carriage crowds, such as 'hostages', 'heroes' and 'seat snatchers' [Davis Associates Ltd, 2005])
3. flow charts (e.g. showing the different stages that workers progress through in their acquisition of the necessary skills or knowledge required to use new technology. It may emphasise potential loopholes or pitfalls along the way.)
4. inventories (e.g. of work pressures)
5. maps (e.g. of the social networks that drivers access when responding to or preparing a response to new technology)
6. matrices (e.g. two-dimensional representations of the work tasks involved in driving a train)
7. scenarios or vignettes (e.g. of realistic situations that could be used during focus groups or interviews or at training days)
8. checklists (e.g. of all the forms of engagement that managers should use to ensure that drivers feel like they have been fully consulted in the selection, design, implementation and use of new technology)
9. lists (e.g. of all the features that drivers do and do not like when given the opportunity to choose between different technologies)
10. statements in a report with the inclusion of units of data as supporting evidence. Most often, these are descriptions of observations or direct quotations from participants. As noted above, quotations should not only convey dominant views across the sample of participants but should also indicate a range of views or exceptions to the dominant view. Whilst counts or percentages are not expected (and are not usually the main point), qualitative phrases such as 'some', 'many', 'a few', 'around half', etc., provide a sense of agreement or disagreement across the sample. Some researchers develop a key for such phrases where 'a few' might indicate between three and five participants or 5–10 per cent.

Summary

Whilst it is useful, if not necessary, to discern between qualitative and quantitative research it is more important to talk about research from a 'fit for purpose' perspective. Being able to design, conduct and interpret any form of research requires an understanding of the best means of answering a particular question. It also requires an acceptance that trade-offs are present in most if not all research, to various degrees. This chapter has provided a greater understanding of qualitative and ethnographic research and its potential use in evaluating new technologies in the rail industry. Finally, technology is inseparable from human designers, implementers and users. Qualitative research is ideally positioned to incorporate these human dimensions in ways that value and preserve the views, experiences and perspectives of real users.

Acknowledgments

This chapter is largely based on research methodology workshops given to Masters Candidates in Safety Management and Clinical Psychology at the University of South Australia from 2008–2011. I am grateful to the students for their involvement and feedback. This chapter is dedicated to all those who have generously participated in and supported my research, with special thanks to Australian train drivers. The chapter has benefitted from the careful and insightful review of Chris Bearman, Anjum Naweed and Drew Dawson.

References

Agar, M. (1973). *Ripping and Running: A Formal Ethnography of Urban Heroin Addicts*. New York: Seminar Press.

Agar, M. (2006). An ethnography by any other name …. *Forum Qualitative Sozialforschung / Forum: Qualitative Social Research* [On-line Journal], 7, [149 paragraphs].

Agar, M.H. (1980). *The Professional Stranger: An Informal Introduction to Ethnography*. New York: Academic Press.

Aull Davies, C. (1999). *Reflexive Ethnography: A Guide to Researching Selves and Others*. London: Routledge.

Baker, S.E., and Edwards, R. (2012). *How Many Qualitative Interviews Is Enough? Expert Voices and Early Career Reflections on Sampling and Cases in Qualitative Research*. UK: National Centre for Research Methods.

Ball, L.J., and Ormerod, T.C. (2000). Putting ethnography to work: the case for a cognitive ethnography of design. *International Journal of Human-Computer Studies*, 53, 147–68.

Becker, H.S. (2007). *Telling About Society*. Chicago and London: University of Chicago Press.

Bernard, H.R. (1995). *Research Methods in Anthropology: Qualitative and Quantitative Approaches*. Lanham, MD: AltaMira Press.

Buchanan, D., Boddy, D., and McCalman, J. (1988). Getting in, getting on, getting out, and getting back. In A. Bryman (ed.), *Doing Research in Organizations*. London: Routledge.

Clarke, A.J. (ed.) (2011). *Design Anthropology: Object Culture in the 21st Century*. New York: SprengerWien.

Coffey, A. (1999). *The Ethnographic Self: Fieldwork and the Representation of Identity*. London: Sage.

Davis Associates Ltd (2005). *Crowd Management on Trains: A Good Practice Guide*. London: Rail Safety and Standards Board.

Edwards, E. (1972). Man and machine: Systems for safety. *Proceedings of British Airline Pilots Associations Technical Symposium* (pp. 21–36). London: British Airline Pilots Associations.

Falzon, M-A., and Hall, C. (eds) (2009). *Multi-Sited Ethnography: Theory, Praxis and Locality in Contemporary Research.* Farnham: Ashgate Publishing.

Gerard Forsey, M. (2010). Ethnography as participant listening. *Ethnography, 11*, 558–72.

Goffman, E. (1959). *The Presentation of Self in Everyday Life*. London: Penguin Books.

Green, J., Willis, K., Hughes, E., Small, R., Welch, N., Gibbs, L., and Daly, J. (2007). Generating best evidence from qualitative research: the role of data analysis. *Australian and New Zealand Journal of Public Health, 31*, 545–50.

Guest, G., Bunce, A., and Johnson, L. (2006). How Many Interviews Are Enough?: An Experiment with Data Saturation and Variability. *Field Methods, 18*, 59–82.

Gurney, J.N. (1985). Not one of the guys: The female researcher in a male-dominated setting. *Journal of Qualitative Sociology, 8*(1), 42–62.

Hawkins, F.H. (1987). *Human Factors in Flight*. Aldershot: Ashgate Publishing.

Hirsch, L., and Thompson, K. (2011a). The Carriage as an Emotional Landscape: How Do Passengers Experience Fear? Understanding Fear in the Australian Metropolitan Railway Industry. *AusRAIL Plus, 22-24 November 2011, Brisbane, Queensland, Australia*.

Hirsch, L., and Thompson, K. (2011b). I can sit but I'd rather stand: commuters' experience of crowding, design and fellow passenger behaviour in carriages on Australian metropolitan trains. *The Australian Transport Research Forum, Hilton Hotel, Adelaide, South Australia, Australia*.

Hirsch, L., and Thompson, K. (2011c). Tarzan travellers: Australian rail passenger perspectives of the design of handholds in carriages. *HFESA 47th Annual Conference 2011, 7–9 November 2011, Crows Nest, NSW.* Baulkham Hills, NSW: The Human Factors and Ergonomics Society of Australia.

Hofstede, G. (1983). *Culture and Management Development*. Geneva: Geneva International Labour Office.

Jolly, L., Radcliffe, D., Smith, A., Nycyk, M., and Andersen, J. (2005/2006). Techno-social systems in organisations. *The International Journal of Technology, Knowledge and Society, 1*, 91–100.

Jordan, B., and Dalal, B. (2006). Persuasive encounters: ethnography in the corporation. *Field Methods, 18*, 359–81.

Keaveney, S.M. (2008). Equines and their human companions. *Journal of Business Research, 61*, 444–54.

Kouabenan, D.R. (2009). Role of beliefs in accident and risk analysis and prevention. *Safety Science, 47*, 767–76.

Malinowski, B. (1961). *Argonauts of the Western Pacific: An Account of Native Enterprise and Adventure in the Archipelagoes of Melanisian New Guinea.* New York: E.P. Dutton.

Mannay, D. (2010). Making the familiar strange: can visual research methods render the familiar setting more perceptible? *Qualitative Research, 10*, 91–111.

McHale, E. (2003). An ethnography of the Saratoga Racetrack. Voices. *The Journal of New York Folklore, 29*, 7–11.

Meyer, D.Z., and Avery, L.M. (2009). Excel as a qualitative data analysis tool. *Field Methods, 21*, 91–112.

Miner, H.M. (1956). Body ritual among the Nacirema. *American Anthropologist, 58*, 503–7.

Morse, J.M. (1999). The armchair walkthrough. *Qualitative Health Research, 9*, 435–6.

Mueller, S., Thompson, K., and Hirsch, L. (2012). *A socio-economic Study of Platform and Carriage Crowding in the Australian Railway Industry: Quantitative Research Summary.* Brisbane, Australia: CRC for Rail Innovation.

Palmer, C., and Thompson, K. (2010). Everyday risks and professional dilemmas: fieldwork with alcohol-based (sporting) subcultures. *Qualitative Research, 10*, 421–40.

Peirano, M.G.S. (1998). When anthropology is at home: the different contexts of a single discipline. *Annual Review of Anthropology, 27*, 105–28.

Rainbird, S., Thompson, K., and Dawson, D. (2010). The impact of organisational culture on fatigue management: the case of camaraderie amongst metropolitan train drivers. *The 7th Annual Meeting of the Australasian Chronobiology Society (ACS)* (pp. 29–33), 4–5 September 2010, Adelaide, South Australia.

Rapley, T.J. (2001). The art(fulness) of open-ended interviewing: some considerations on analysing interviews. *Qualitative Research, 1*, 303–23.

Richards, L. (2005). *Handling Qualitative Data: A Practical Guide.* London: SAGE Publications.

Roth, W-M. (2006). But does "ethnography by any other name" really promote real ethnography? *Forum Qualitative Sozialforschung / Forum: Qualitative Social Research [On-line Journal], 7*, [33 paragraphs].

Saldana, J. (2009). *The Coding Manual for Qualitative Researchers*. London: SAGE publications.

Sale, J.E.M., Lohfeld, L.H., and Brazil, K. (2002). Revisiting the quantitative-qualitative debate: implications for mixed-methods research. Quality &amp. *Quantity*, *36*, 43–53.

Sands, R.R. (1999). Experiential ethnography: playing with the boys. In R.R. Sands (ed.), *Anthropology, Sport, and Culture*. Westport, CT: Bergin and Garvey.

Savage, J. (2006). Ethnographic evidence: The value of applied ethnography in healthcare. *Journal of Research in Nursing*, *11*, 383–93.

Shank, G.D. (2002). *Qualitative Research: A Personal Skills Approach*. Upper Saddle River, NJ: Merrill Prentice Hall.

Spradley, J. (1980). *Participant Observation*. New York: Holt, Rinehart and Winston, Inc.

Steier, F. (ed.) (1991). *Research and Reflexivity*. London: Sage.

Stiles, M.C., and Grieshop, J.I. (1999). Impacts of culture on driver knowledge and safety device usage among Hispanic farm workers. *Accident Analysis & Prevention*, *31*, 235–41.

Thomas, W.I., and Thomas, D.S. (1928). *The Child in America: Behaviour Problems and Programs*. New York: Alfred A. Knopf.

Thompson, K. (2004). La Marismeña Australiana. In J. Fook, S. Hawthorne and R. Klein (eds.), *Horse Dreams: The Meaning of Horses in Women's Lives*. Melbourne: Spinifex Press.

Thompson, K. (2007a). Le voyage du centaure: la monte à la lance en espagne (xive–xxie siècles). In D. Roche and D. Reytier (eds.), *À cheval! Écuyers, amazones & cavaliers du XIVe au XXIe siècle*. Paris: Association pour l'Académie d'Art Équestre de Versailles.

Thompson, K. (2007b). *Performing human-animal relations in Spain: An anthropological study of the bullfighting from horseback in Andalusia*. (Doctoral thesis). University of Adelaide, Adelaide, South Australia.

Thompson, K. (2010a). Binaries, boundaries and bullfighting: multiple and alternative human-animal relations in the Spanish mounted bullfight. *Anthrozoos: A Multidisciplinary Journal of the Interactions of People and Animals*, *23*, 317–36.

Thompson, K. (2010b). Narratives of tradition: The invention of mounted bullfighting (rejoneo) as "the newest but also the oldest". *Social Science History*, *34*, 523–61.

Thompson, K. (2011). Theorising rider-horse relations: An ethnographic illustration of the centaur metaphor in the Spanish bullfight. In N. Taylor and T. Signal (eds.), *Theorising Animals*. Leiden, Boston: Brill.

Thompson, K. (2013). From Initiate to Insider: Renegotiating Workplace Roles and Relations Using Staged Humorous Events. *Organization Management Journal*, *10*(2), 122–38.

Thompson, K., and Hirsch, L. (2011). *A Socio-economic Study of Carriage and Platform Crowding in the Australian Railway Industry: Implications for Further Research.* Brisbane: CRC for Rail Innovation.

Thompson, K., Hirsch, L., Mueller, S., Sharma-Brymer, V., Rainbird, S., Titchener, K., Thomas, M., and Dawson, D. (2012). Riding a mile in their shoes: Understanding Australian metropolitan rail passenger perceptions and experiences of crowdedness using mixed-methods research. *Road and Transport Research Journal, 21,* 46–59.

Thompson, K., Hirsch, L., Mueller, S., and Rainbird, S. (2012). *A Socio-economic Study of Carriage and Platform Crowding in the Australian Railway Industry: Final Report.* Brisbane: CRC for Rail Innovation.

Thompson, K., Hirsch, L., and Rainbird, S. (2011). *A Socio-economic Study of Carriage and Platform Crowding in the Australian Railway Industry: Qualitative Research Summary.* Brisbane: CRC for Rail Innovation.

Thompson, K., Palmer, C., and Raven, M. (2006). *Alcohol in the Lives of Australian Rules Football Fans: Social Meanings and Public Health Implications.* Flinders University, Adelaide: Department of Public Health.

Thompson, K., Palmer, C., and Raven, M. (2011). Drinkers, non-drinkers and deferrers: Reconsidering the beer/footy couplet amongst Australian Rules football fans. *The Australian Journal of Anthropology, 22,* 388–408.

Thompson, K., Pirone, C., Crotty, M., Hancock, E., Thomas, M., and Turner, P. (2010). *SafeTECH: Safe Tools for Electronic Clinical Handover – Summary of Research Findings.* Adelaide: South Australian Department of Health.

Thompson, K., Rainbird, S., and Dawson, D. (2010). "The nature of the beast": Metropolitan train drivers' experience, perception and recognition of fatigue. *Sleep, Circadian Rhythms and Health: Proceedings of the 7th Annual Meeting of the Australasian Chronobiology Society* (pp. 1–5). Adelaide, South Australia: Australasian Chronobiology Society.

Thurmond, V.A. 2001. The Point of Triangulation. *Journal of Nursing Scholarship, 33,* 253–8.

Wilson, S., Galliers, J., and Fone, J. (2007). Cognitive Artifacts in Support of Medical Shift Handover: An In Use, In Situ Evaluation. *International Journal of Human-Computer Interaction, 22,* 59–80.

Zimmermann, B. (2007). Tracing the action of technical objects in an ethnography: vinyls in Beijing. *Qualitative Sociology Review, 3*(3), 22–45.

Chapter 6

Future Inquiry: A Participatory Ergonomics Approach to Evaluating New Technology

Verna Blewett

Central Queensland University, Appleton Institute, Adelaide, Australia

Andrea Shaw

Shaw Idea Pty Ltd, Mount Egerton, Victoria, Australia

Introduction

How is it possible to get agreement to move forward when a community, industry or organisation presents with conflicting opinions, all held equally dearly? How can new technologies be evaluated in such a way that many voices can be heard? How can divergent ideas for change be translated into action that makes a difference? While the importance of stakeholder involvement in effective organisational change has long been recognised (Lewin, 1952) and there are numerous processes for engagement available in the marketplace (Holman and Devane, 1999), techniques that enable workable outcomes in the context of conflict and disagreement are not common. Many of the available techniques fail to effectively include different interests and thus do not support true engagement. They can more accurately be characterised as consultative (often interpreted as management informing workers of change) rather than participative (in which many voices take part in the development of change) since they do not provide less powerful stakeholders with real influence over outcomes. This partly results from reluctance to deal with conflict in organisational change processes; unrealistic performance targets, short timelines, historic conflicts and limited experience of participation can all result in processes that seek to avoid confrontation, rather than using different points of view to engender a creative response.

Organisations and industries necessarily consist of groups of actors with different opinions and ideas that create a complex political environment. In such environments processes that do not recognise and deal with disagreement and conflict can lead to the breakdown of key relationships that are, in fact, necessary for success. For many organisations, organisational politics becomes a key contextual constraint in the development and implementation of effective change strategy, rather than a source of energy and ideas for change (Buchanan and Dawson, 2007). Instead of constraining change, an effective participative process can harness the different points of view within and surrounding an organisation to provide complete information and a full range of options for change. As well as this instrumental purpose, a truly participative process treats all of the people involved with dignity and respect. This chapter describes one such process, called 'Future Inquiry', that we have developed from the principles of Appreciative Inquiry (Cooperrider, Whitney and Stavros, 2003) and Future Search (Weisbord and Janoff, 2000, 2010), which enables diverse groups to work together to determine an agreed agenda for change.

Future Inquiry embodies the principles of participation and respect that underpin effective participatory processes, ensuring that these are built in from the beginning. Because it takes a human-centred approach, we classify it in the domain of participatory ergonomics. The Future Inquiry process occurs through a workshop which engages a large group of people who are representative of the community, industry or organisation, and also representative of the particular focus of the activity, that is, the inquiry. We actively seek common ground as a basis for action since people are more likely to take action where they have shared interest and commitment. Having taken action on common ground, groups are likely to find that other, sometimes more difficult, areas can subsequently be tackled. We focus on the future rather than on problem-solving past errors, and provide an environment in which participants manage their time and discussions within a directed framework of activities.

This chapter describes the Future Inquiry workshop, including an example of its use in the *Keeping Rail on Track* research into organisational culture and work health and safety in the Australian rail industry (Blewett, Rainbird, Paterson, Etherton and Dorrian, 2012). *Keeping Rail on Track* used a multi-method approach to investigate organisational culture and its relationship to work health and safety in three rail sectors: national freight, heavy haul and urban passenger rail services. It used a mix of quantitative and qualitative methods and used the findings of prior research in the Australian mining industry, *Digging Deeper* (Shaw et al., 2008), as an analytical framework. Following preliminary data analysis and examination of key literature, the first author facilitated a Future Inquiry workshop in the national freight rail firm to identify potential intervention strategies aimed at improving work health and safety.

Background

Participatory ergonomics has a relatively recent history in the domain of organisational change but has rapidly growing acceptance as the *modus operandi* for ergonomists and human factors professionals seeking effective interventions to improve the working environment. With roots in the work of Lewin (1946) who identified the importance of worker involvement, and Emery and Trist (1973) who advocated a view of organisations as socio-technical systems, participatory ergonomics is not a prescribed process but rather a set of parameters that the process should meet. As Imada (1991) defined it, '... participatory ergonomics requires that the end user (the beneficiaries of ergonomics) be vitally involved in developing and implementing the technology' (p. 32).

Nagamachi (1995) recommended that the workers' active involvement should be supported by their supervisors and managers and developed in a team-based manner to apply ergonomics principles to the problem at hand. More recently, Haines, Wilson, Vink and Konigsveld (2002) described a framework for characterising participative ergonomics interventions that has nine dimensions: permanence, involvement, level of influence, decision-making power, mix of participants, requirement, focus, remit, and role of the ergonomics/human factors specialist. The authors suggest that the complexity of a project is likely to define the types of participatory mechanisms that are used (e.g. small groups, groups across departments, multi-level groups) but that the nine dimensions will apply in any case. The approach taken in participative ergonomics interventions can have a variety of features in each of these dimensions but they suggest that fundamentally it must provide the participants in the process with control over the outcomes if participation is to be legitimate rather than token.

In terms of evaluating new technology, the dimensions could be applied to a participative ergonomics task force as follows:

- Permanence: There may be a temporary task force for a one-off evaluation, or if the evaluation is to be ongoing and iterative, then a permanent group might be established.
- Involvement: Some key people may be chosen for the task force because of their knowledge or level of influence, or members may be there to represent particular sectional interests, or there may be some mix of the two.
- Level of influence: It may be that the task force is confined to the area affected by the new technology, or it may be across departments or the whole organisation.
- Decision-making power: This dimension identifies who in the group has the power to make decisions; it may remain with someone outside the task force.
- Mix of participants: The categories needed for effective participation are considered in this dimension, for example, operators, supervisors, specialist roles and management people.
- Requirement: Involvement in a participatory ergonomics task force may be specified as part of the job or it may be voluntary.
- Focus: This dimension defines the role of the task force and what it is formed to do.
- Remit: Identifies the activities of the task force and its level of influence and involvement in change; particularly important if the task force is permanent and coordinates with other continuous improvement activities.
- Finally, the role of the ergonomics/human factors specialist is described so that the facilitative role is clear but allows for evolution over time.

There is no single, perfect technique for developing such interventions and seeking their implementation, no neat one-size-fits-all approach that comes with guaranteed success. In fact there are many processes that work, some of them well documented and accessible (Holman and Devane, 1999), and what makes them work may not be the intrinsic design of the process but factors to do with the group, its commitment to participative change and the skill of the facilitator in guiding, preparing and conducting the process (Weisbord, 2008). However, no matter how well-intentioned a program of change is, without appreciating the politics of implementation at organisation or industry level, initiatives are likely to fail (Blewett, 1999). Thus, a process leading to change must take internal politics into account as part of the landscape of change (Buchanan and Badham, 1999), not just as potential impediments, constraints or enablers, without allowing them to take over the agenda. Two participatory planning approaches have demonstrated great capacity to achieve this and lead to durable change.

The first approach is Appreciative Inquiry (Whitney and Cooperrider, 1998). Appreciative Inquiry focuses on examining new directions for action by looking at what works well now and building on it, rather than problem-solving. Problem-solving tends to be slow and invites examining the past to look for the causes of problems. It can be limited to closing gaps rather than looking for expansive, fresh ideas and it tends to generate defensiveness (by blaming others for the problems, rather than self-reflexivity) thus reinforcing power and control agendas. Appreciative Inquiry brings a focus on positive stories and ideas that aims to generate respect for what has been done well, identifies the part that individuals play in their organisation, reinforces accepted values and invites an affirmation of ideas. Cooperrider et al. (2003) suggest that organisations tend to move toward what they study, therefore a focus on the positive as a blueprint for the future should lead organisations to improvement. However, we find that denying attention on what does not work can frustrate people who wish to have their grievances acknowledged as a first step to engagement or it can generate 'unjustified and intemperate optimism' (Rogers and Fraser, 2003, p. 77), leading to unrealistic change strategies. We also find that the structure of Appreciative Inquiry can lack the capacity to generate plans for action.

The second approach is Future Search (Weisbord and Janoff, 2000; Weisbord and Janoff, 2010), a collaborative large-group process that involves 'the whole system', that is, participants from all parts of the system are involved. The aim is to make sure that the voices of the whole system are in the room. Using a structured and focused agenda that examines the past, the present and the future, the many perspectives of the participants are brought together to identify 'common ground'. The process encourages creativity and commitment to actions that are grounded in reality; the outcomes are more likely to be implemented because the process identifies actions to which people are prepared to commit (Cialdini, 2001; Weisbord and Janoff, 2005). We have also observed voluntary cooperation and the formation of new working relationships between people who were traditionally politically opposed. Participants share leadership and engage as peers in robust discussion, in an environment that is focused on the future. If the right people are in the room and it is well-planned and expertly facilitated, then time and resources spent on a Future Search can be well spent. But often a client company or industry group will not be able to commit 30–100 people for the required 16 hours over three days.

Faced with the dilemma of needing to engage the whole system, find common ground and appreciate what works and build on it, and do it in a maximum of one day, we developed Future Inquiry by amalgamating these two key approaches with some of our own techniques. Future Inquiry engages people in thinking past 'problems' to thinking about a desired future that they are willing to work towards. Many of the techniques that we use are modelled on Future Search techniques but modified to enable the whole workshop to be conducted within one day. For example, in designing the future we define a specific task with constrained rules that can be conducted in 30–40 minutes. In contrast, Future Search includes a highly creative exercise for designing the future; it takes several hours to complete but the results are exhaustive, detailed and descriptive.

We do not see Future Inquiry as a substitute for Appreciative Inquiry or Future Search but it is a useful alternative to have in the facilitator's toolbox when circumstances demand it. For the case study presented here, the logistics involved in bringing all of the stakeholders together from across the organisation was only feasible for a one-day workshop. There was also known conflict between some of the stakeholder groups that needed to be carefully managed, thus we considered Future Inquiry, with its egalitarian principles and capacity to provide a 'safe' environment, to be the most suitable participative planning tool. We also used the Future Inquiry workshop to collect data on the different perceptions of the research within the organisation, to hear the organisation's views of the ideal future and to discuss options for the implementation of strategies for change that would lead to the desired future. These data informed the final report and its recommendations (Blewett et al., 2012).

Setting up a Future Inquiry Workshop

Planning is crucial to the success of a Future Inquiry workshop and requires detailed knowledge of the system concerned. In our example, the first author worked with a tripartite commissioning group (the Steering Group) that included representatives from management and labour. In other instances, particularly when working with an industry group, a tripartite steering group that covers the key industrial, organisational and government groups in the system may be appropriate. The primary need is to ensure that the right people are identified and recruited to be in the room for the Future Inquiry workshop. Getting the whole system in the room can be a very discomforting proposition for some organisational actors where significant conflict exists or where there is a strong belief that the views of different groups are irreconcilable. Providing a facilitated venue for the steering group to talk about the process is a first step to building confidence within the group. The steering group defines the topic of the Future Inquiry workshop and establishes any boundaries for discussion that might exist. We define our needs for the workshop logistics (as facilitators) but leave as much of the management and organisation of the workshop to the steering group as we can, in order to build commitment to making the workshop and its outcomes a success.

During the planning stages we ask the steering group to identify the stakeholders that have influence over or who are affected by the issue under discussion and who therefore should be considered for participation in the workshop. In particular, we encourage the steering group to identify 'outsiders'; groups or individuals with influence over the issue but who are usually considered to be outside the system, for example regulators, contractors, customers, suppliers, advisors. We brainstorm an initial list of individuals and groups and then refine this list to form about eight key stakeholder groups. We identify someone from each stakeholder group with whom to liaise in order to arrange attendance. The responsibility for determining which specific individuals attend the Future Inquiry workshop is usually passed to each stakeholder group to determine for itself.

In our example, the Steering Group consisted of the organisation's occupational health and safety committee plus the finance manager and an organiser from the rail union (RBTU). They selected six stakeholder groups: senior management, executive management, front-line management, workers (including contractors), corporate, and worker representatives (the unions and health and safety representatives). Each group selected five or six members, except for the worker group which had nine members. A total of 38 participants attended the day-long workshop.

Low Technology

We prefer to use low technology during a Future Inquiry workshop because it is reliable, readily available, inexpensive and effective. We use flip charts and easels, water-based marker pens, masking tape and removable adhesive gum. We ask people to write their discussions, observations, diagrams and drawings on flip charts that are then displayed on the walls for everyone else to see and so that they are available throughout the day. Participants wear name badges with no titles but with their preferred given name written in large print and their family name and affiliation (if needed) written in small print so that traditional hierarchies are challenged from the start. Membership of stakeholder groups is identified with a coloured sticker on each name badge. During the day, participants are directed into stakeholder groups by colour and into mixed groups by asking groups to make sure that at least one of each colour is represented. We sometimes pre-determine the membership of mixed groups by using a number on the name badge to direct participants into groups to avoid collusion between known influential participants but more usually, and in this example, we allow mixed groups to be self-selected. We do not provide tables for participants, instead participants move their chairs to form and re-form groups of various sizes and memberships, depending on the exercise. We find this enhances the discussion in small groups because, with no table to separate people, they are likely to remain engaged. We ask for a square room that will take all the participants in a single-row circle and use this configuration for most plenary sessions so that everyone can see others in the room. We prefer natural light sources and ask that healthy food is served during the breaks so that people are given the best chance to keep focused and alert. We need to be sure that the venue managers will allow us to either stick flip chart paper to the walls or provide us with sufficient display surfaces. We call these features our 'conditions for success' drawing from Future Search (Weisbord and Janoff, 2000; Weisbord and Janoff, 2010) and confirmed by our own experience.

Facilitation

If the group is larger than 40 participants we use two facilitators because the role is demanding; there are many participants and there is generally a lot of activity in the room. Two facilitators can better maintain the energy in the room by swapping tasks and allowing each other to rest briefly. This tandem activity requires skill and considerable trust between the facilitators. Experience working together enables quick changes to the program or activities in response to events in the room without the rhythm of the day being disturbed for the participants.

In our case study, the first author was the sole facilitator as well as a content expert and researcher. Although we use Future Inquiry in areas where we have only limited knowledge of the content of the inquiry, we are frequently called to be both facilitators and content experts. In these cases we make a clear distinction between our contribution as content experts during the first session of the workshop (presenting the past) and our role as facilitators in subsequent sessions. This is not an impediment to facilitation because we genuinely want to hear what the group is preparing to commit to doing as a result of the research or consultancy activity and we trust the process to give us this information. We demonstrate this desire by standing back and giving the floor to the participants, not judging their input or 'correcting' their statements. What we learn is used to inform our recommendations not only because of the content that participants offer but also because we gain an understanding of the political context and the limitations on implementation that may exist.

The Future Inquiry Workshop Format

The Future Inquiry workshop we describe here was an organisation-wide exploration of how it might better manage and incorporate work health and safety into the strategic direction of the company. Such change involved the incorporation of new technology into the workplace and the evaluation of those technologies: specifically, new computer systems, new infrastructure, new systems for managing rosters and fatigue at work. Each of the Future Inquiry workshops we have facilitated has had a similar format, although the workshop tasks vary with each assignment. Variation arises because of the nature of the matter being investigated, the nature of the group that will be involved and the time available for the workshop; sometimes the 'day' may be as little as 10:00 am to 4:00 pm. Drawing on Future Search, we move participants from an examination of the past, to the present and then to the future with respect to the topic. Participants work in stakeholder groups for some exercises and in mixed groups for others, giving them the opportunity to test ideas with their peers, especially in the first stages of the workshop, and at other times representing their stakeholder voice with people from other stakeholder groups. Some exercises are done as a plenary group so that the whole room is able to hear the different views that exist within the whole system.

The Past

In organisation-level or research-oriented Future Inquiry workshops, such as our example, we typically provide a brief report on the topic to participants for reading in advance of the workshop, covering the findings of a literature review or preliminary research findings. We then review this in the first session of the workshop, providing a short, formal presentation in order to answer the question, 'What do we know about the topic already?' This is particularly important in the research context because an extensive literature review is generally part of the research project and informs the research method, decisions about data analysis and recommendations for action. We have found that presentation of a literature review to the whole system in the room can be a 'reality check'; an opportunity to dispel some myths, and establish a baseline or boundary for discussion. We field questions from the participants at this point to make sure that there is shared understanding in the room. The remainder of the workshop is handed to the participants while we withdraw as content experts and act as facilitators. In our example, the first author presented the findings of the first stage of data collection in the organisation: a survey, focus groups and interviews with a cross-section of the organisation. We reminded participants that some months had elapsed since the data were collected, thus the findings were genuinely a representation of the past because the organisation had continued to change and grow in the intervening period. The presentation allowed boundaries to be set around the breadth of discussion and participants were advised that their input would not *form* the recommendations and final report but that they would *inform* them.

The Present

We gain an appreciation of the present by asking participants to work in stakeholder groups to identify what works now and what does not work now in relation to the topic of the workshop. This process draws on Appreciative Inquiry. The ideas are collected on flip charts and hung on one wall for everyone to review during the morning tea break. Doing this review task during a natural break in formal proceedings invites people to talk casually. Common themes can be readily seen by this stage and we overhear comments from members of traditionally opposed groups such as, 'I didn't know you held the same view as us ....' The common themes are collected on flip charts during a plenary session immediately after the break and these too are displayed on the wall.

In our example, participants commonly identified that safety was regarded as the number one priority across the organisation, that training and mentoring were strengths and the culture of open communication was appreciated. Participants commonly identified that insufficient time was allocated to deal with issues, that some people demonstrated negative attitudes and took an 'I don't care' approach, and that in general the organisation was reactionary – seeking to mop up after the event rather than fix things to prevent problems. Some of these issues had been contested territory in discussions within the organisation, so it was significant to see the industrial players reaching agreement in this setting.

The next activity is to scan the current environment by discussing trends that influence the capacity for the industry or organisation to do well right now in the area under discussion. The whole group is involved in this and develops a large, group mind map. The centre of the mind map contains the topic area expressed as the ideal, as in our example 'the organisation's great safety culture'. The trends collected on the mind map have different priority for different people and different groups of stakeholders. Each participant is given five sticky dots in their stakeholder colour and is invited to 'vote' for the trends that they consider the most important or significant. They spread their dots across the mind map as they choose.

In our example, key trends included organisation-level observations as well as industry-level trends: the family-oriented nature of the organisation, and the increasing identity of the organisation in the rail industry at the organisational level, and the increasing regulatory burden in the industry, and changing demographics of the rail industry at the industry level.

After voting, stakeholder groups then re-convene to examine the voting on the mind map and select a few trends considered significant by their group. They are asked to identify what they (as individuals or as a stakeholder group) are doing now in response to those trends and what they *could* do now in response to those trends but are not doing. We stress that this is an opportunity for self-reflection to examine what the stakeholder group itself may have done and not done, not an opportunity to lay blame at the feet of others for things that have or have not been done. Thus they create a clear picture of the present that is shared by others in the room. Gaps in the present become clear and a sense of uncertainty allows the group to move to consideration of the future.

The Future

Having discussed the past and examined the present as stakeholders and as a plenary group, we turn to the future. We ask participants to project themselves

10 years into the future (or whatever period makes sense to the group) and use the present tense to describe that future. The change in language is necessary to shift participants' thinking so they 'look back from the future' rather than 'look towards the future'. They are to consider that all problems concerning the topic under consideration have been solved and the situation is ideal. Mixed groups, consisting of at least one member from each stakeholder group, assemble to describe this ideal future and write the front page of the organisation's newsletter, dated 10 years in the future (that is, the projected 'present'), describing what has been achieved to create this ideal future and the work done since the workshop (10 years ago) that enabled them to reach that future. They are asked to consider questions such as: What barriers did you overcome back in the year of the workshop to get to today? How did you do it? What is being delivered for the industry? For your organisation? For people who work in the industry? For families and the community? For the economy? For the nation? They are asked to include in their description a headline, an image, a list of key milestones along the way and a quote from each group member about how they feel about the changes over the last 10 years. They are given boundaries on their desired futures by being asked to make sure that what they describe is feasible (people could do it if they wish), desirable (the whole community would benefit) and motivating (they would be willing to work to make it happen). The completed newsletters are hung on the wall and the whole group examines them, looking for similarities and differences.

In our case study Future Inquiry workshop, the first author asked participants to imagine a future for the organisation where the systems and features of the organisation were in place to successfully lead to a wonderful safety culture. Groups identified themes dealing with both processes used and outcomes evident in their ideal future.

Getting to the Future

After examining the different visions of the ideal future, the whole group gathers in front of the newsletters to discuss the areas of 'common ground' they contain, that is, to identify the common themes contained in the different groups' descriptions of their ideal future. This helps to determine what people commonly want and what they would be prepared to work for. These items are recorded on flip charts and displayed on the wall. The common ground identified from this process in our example included: acknowledgement of the evident commitment to the organisation, the sense of fun that pervaded the organisation, the desire to use modern systems to increase railcar driver retention and improve training, the shared desire to improve environmental management, the shared focus on improvement strategies, and the involvement of everyone.

Moving the group from the pleasant place of the ideal future to the reality of the need to work to get there is softened by focusing participants on the steps taken to get to the desired future. While still 'in the future' participants prepare a 'reverse history' that describes in detail how they moved from the date of the workshop to the present. They do this individually or in small groups (as they choose) by writing key events, strategies or people's actions on sticky notes. They are asked to write one item only on each sticky note and also write the year it happened. The sticky notes are then put onto the wall by year – a place for each of the preceding 10 years. Participants group the sticky notes in each year by the common themes. This exercise transforms the future as a theoretical exercise into something concrete as they identify 'actual events' that have occurred in the imagined past. At the end of the exercise, the group is brought back to the real present and the next stage is introduced.

In a plenary discussion, common themes from the newsletters and the reverse history are worded to form strategies for action that would enable movement towards the desired future as outlined in the newsletters. The strategies are divided amongst the plenary group with self-selected groups working on specific strategies that the whole group identified as having priority. In Future Search terms, this is the 'common ground'. Each group decides what should be done to implement the strategy, what important outcomes are expected, what information or data are needed, what drivers and barriers to change exist and what needs to be done. Most importantly, they identify the 'do-able' first steps towards the desired future and who should take them. Short reports are made to the whole group and we encourage someone in the room (not necessarily in the same sub-group) to take carriage of that first, do-able step. This process provides the opportunity for commitment and consistency, public announcement (within the group) and approval for action from the group (Cialdini, 2001). The workshop ends with the whole group taking a few minutes to reflect on the day and the actions they commit to taking on the following working day to move the collective agenda forward. In our example, the workshop ended with commitment by the participants to collecting data and taking action to improve environmental management, leadership by management and employees, consultation and communication, the safe growth of the company and improved recruitment and conditions of employment. Reflecting on the workshop the following day, one senior manager told us 'The most important lesson for management was finding that the employees are committed to the future development of the company and are eager to have a voice'.

Conclusion: Using Future Inquiry

We have used Future Inquiry in a variety of settings and have found it particularly useful as a means of evaluating new technologies, as data collection for research, in developing an industry-based strategy and in the first stages of planning in an organisation or community group. It is a participative process that allows differences between stakeholders to be acknowledged without causing conflict. It also provides the opportunity for participants to identify an agreed path forward. While it does not have the power of a full Future Search conference, it is built on the same values and establishes respectful, egalitarian relationships from the outset.

It also applies the lessons of Appreciative Inquiry by focusing on positive stories and ideas, which generates respect for what has been done well, identifies the parts that individuals play in their organisations, articulates accepted values and invites an affirmation and expansion of ideas. The Future Inquiry workshop yields insights that are grounded in the experience of stakeholders, reflecting the reality of everyday working life and identifying existing strengths as well as needs. It provides a format for not only determining a desirable and feasible future but also determining the first steps needed to get there. As an evaluative tool it draws on the current and past experiences of a variety of stakeholders which enables new technologies to be contextualised into the reality of everyday working life in order to answer questions about how new technologies might fit or alter current activities. This participative approach to the evaluation of new technologies encourages the consideration of many and varied voices, providing a complex, all-round examination of proposals for change in an efficient time-frame.

Acknowledgements

The authors are grateful to the CRC for Rail Innovation (established and supported under the Australian Government's Cooperative Research Centres program) for the funding of this research. Project No. R2.101 Project Title: *Keeping Rail on Track*. We acknowledge the case study organisation's assistance in allowing researchers access to sites, personnel and data, and we are grateful to the organisation's employees and contractors for their willing participation in this research. Thanks to Marvin Weisbord and Dr Sandra Janoff for providing useful comment on an early draft of this chapter.

References

Blewett, V. (1999). Ron Cumming Memorial Lecture: Death, taxes, change and controversy. In Summers, A., Chew, S., Gibson, I., Miller, J. and Pietrocola, W., *Proceedings: Ergonomics Society of Australia Conference: Better Skills, Better Future: Practical Skills for Work, 11–13 October 1999, Fremantle, Western Australia.*

Blewett, V., Rainbird, S., Paterson, J., Etherton, H., and Dorrian, J. (2012). *Keeping Rail on Track*. Brisbane: Central Queensland University, Appleton Institute.

Buchanan, D., and Badham, R. (1999). *Power, Politics, and Organizational Change: Winning the Turf Game*. London: Sage.

Buchanan, D., and Dawson, P. (2007). Discourse and audience: organizational change as multi-story process. *Journal of Management Studies*, *44*(5), 669–86.

Cialdini, R.B. (2001). *Influence: Science and Practice* (4th edn). Boston: Allyn and Bacon.

Cooperrider, D.L., Whitney, D., and Stavros, J.M. (2003). *Appreciative Inquiry Handbook*. Bedford Heights, Ohio: Lakeshore Publishers.

Emery, F.E., and Trist, E.L. (1973). *Towards a Social Ecology: Contextual Appreciations of the Future in the Present*. New York: Plenum Press.

Haines, H., Wilson, J.R., Vink, P., and Konigsveld, E.A.P. (2002). Validating a framework for participatory ergonomics (the PEF). *Ergonomics*, *45*(4), 309–27.

Holman, P., and Devane, T. (eds). (1999). *The Change Handbook: Group Methods for Shaping the Future*. San Francisco: Berrett-Koehler Publishers, Inc.

Imada, A.S. (1991). The rationale and tools of participative ergonomics. In K. Noro and A.S. Imada (eds.), *Participatory Ergonomics* (pp. 30–49). London: Taylor & Francis.

Lewin, K. (1946). Action research and minority problems. *The Journal of Social Issues*, *2*(4), 34–46.

Lewin, K. (1952). *Field Theory in Social Science: Selected Theoretical Papers*. London: Tavistock Publications.

Nagamachi, M. (1995). Requisites and practices of participatory ergonomics. *International Journal of Industrial Ergonomics*, *15*, 371–7.

Rogers, P.J., and Fraser, D. (2003). Appreciating appreciative inquiry. *New Directions for Evaluation*, *100*(Winter), 75–83.

Shaw, A., Blewett, V., Stiller, L., Aickin, C., Cox, S., Ferguson, S., and Frick, K. (2008). Digging Deeper: Wran Consultancy Project Final Report, Vol. 2. Sydney: NSW Mine Safety Advisory Council.

Weisbord, M., and Janoff, S. (2005). Faster, shorter, cheaper may be simple; it's never easy. *Journal of Applied Behavioral Science*, *41*, 70–82.

Weisbord, M.R. (2008). Large groups methods: a three-part shopper's guide. In E. Biech (ed.), *ASTD Handbook for Workplace Learning*. San Francisco: ASTD.

Weisbord, M.R., and Janoff, S. (2000). *Future Search: An Action Guide to Finding Common Ground in Organizations and Communities (2nd edn)*. San Francisco: Berrett-Koehler.

Weisbord, M.R., and Janoff, S. (2010). *Future Search: Getting the Whole System in the Room for Vision, Commitment, and Action (3rd edn)*. San Francisco: Berrett-Koehler.

Whitney, D., and Cooperrider, D. (1998). The appreciative inquiry summit: overview and applications. *Employment Relations Today*, *25*(2), 17–28.

Chapter 7

Using Task Analysis to Inform the Development and Evaluation of New Technologies

Janette Rose
University of South Australia, Adelaide, Australia

Chris Bearman
Central Queensland University, Appleton Institute, Adelaide, Australia

Anjum Naweed
Central Queensland University, Appleton Institute, Adelaide, Australia

Introduction

The increasing introduction of technology into the rail industry has the potential to improve safety and efficiency, however, introducing technology is not without risk and not all technology provides its promised benefits. It is essential, therefore, to thoroughly evaluate a new technology from the perspective of applicability and safety, to ensure that the technology will do what it purports to do and will not create safety risks. One of the steps in this evaluation process is the construction of a task analysis.

Task analysis is a way of conceptualising the tasks that must be completed for a person to be able to perform a given activity. This conceptualisation of the task allows the analyst to closely examine the components of a given activity, consider whether any improvements can be made, and assess the impact of any changes to that activity. Task analysis may be thought of as a method of obtaining user feedback since its construction relies on the input of users, and the results of its application to evaluate a technology should be discussed with both users and experts to ensure that the analysis is accurate.

Task analysis has been used extensively in many industries such as aviation, air traffic control, automotive, product design and military operations (Liljegren, 2006; Stanton, Salmon, Walker, Baber and Jenkins, 2005) and although not as widely used in the rail industry, it has been employed for a variety of different purposes (Tichon, 2007), some of which will be discussed in this chapter.

This chapter begins with a brief history of task analysis from its first conception in the early 1900s ('A Brief History of Task Analysis'). We then explain what a task analysis actually is ('What is Task Analysis') and describe the types of task analyses that are most likely to be of interest in the rail industry, including examples from other domains ('Different Types of Task Analysis'). This is followed by examples of where task analyses have been used for various purposes in the rail industry ('Examples of Task Analyses Conducted in the Rail Industry'). Next, we describe the process undertaken by Rose and Bearman (2012) to compile a goal-directed task analysis and follow this up with some guidelines on choosing the correct method ('Constructing a Task Analysis'). Finally, we show how the goal-directed task analysis can be used to evaluate an in-cab information device ('Example of Applying Task Analysis to Evaluate a New Technology').

A Brief History of Task Analysis

Modern task analysis originated from the scientific management movement of the early 1900s when Taylor (1911) timed workers on various tasks and developed ways of improving performance and reducing accidents for the benefit of the company and the workers. At this time, Taylor looked at simple tasks including loading pig-iron onto railcars, shovelling ore and bricklaying. By observing the workers and conducting experiments, he was able to find the best way to perform a certain task and to calculate how much work could reasonably be completed during a shift. Taylor then looked at approaches to management and developed ways of providing workers with the incentive to work to their level of ability.

Over the following decades the analysis of job design became more frequent and by the 1950s had become an important part of the newly emerged field of ergonomics (Crystal and Ellington, 2004; Stanton, 2006). In the 1960s the forms of task analysis that are present today began to emerge. Using a combination of psychology and engineering principles to break down tasks into their component physical and mental elements, psychologists such as John Annett focused on understanding and improving performance, identifying problems in various stages of task performance and proposing potential solutions (Annett, 2003; Rose and Bearman, 2012). Several of the modern task analysis methods utilised today, such as the hierarchical task analysis, were created in this decade (Stanton et al., 2005).

Since the 1960s, in conjunction with advances in technology, the complexity of task performance has substantially increased and task analysis methods have evolved accordingly (Stanton et al., 2005). One of the ways in which task analysis has evolved is into the mental (or cognitive) realm of the person, so that the task analysis can capture the mental activities that are required, the information a person needs and the decisions that they make. These cognitive factors of a task are particularly important when we consider the human factors implications of a new technology.

What is Task Analysis

Task analysis is a set of slightly different methods that provide a process for identifying exactly what a task is, collecting data on that task, analysing the data so that the task is clearly understood and then producing a documented representation of the analysed task (Annett, Duncan, Stammers and Gray, 1971). In most cases, task analyses are used to improve understanding of human-machine and human-human work-based interactions by breaking down tasks into their component steps and the operations required to perform those steps (Stanton, et al., 2005). In this way, a better understanding of the task can be developed, interventions can be designed to improve performance and changes to the task itself or the environment in which the task occurs can be evaluated.

The procedure for most task analyses is very similar. They generally begin with defining the task and/or purpose of the analysis. In some cases collecting data precedes definition of the overall goal and sub-goals, while in other cases the collection of data follows the definition of goals. Collection of data uses various techniques, such as interviews and observations. Once data is collected, the next step is usually analysis or organisation of the data, and this is followed by an output such as hypothetical solutions to problems.

Task analysis can be used in different ways to evaluate a new technology. In its most basic form, a task analysis essentially provides a structured approach to thinking through the ways that introducing new technology will change specific tasks that make up an operator's role. At a more sophisticated level, task analysis can be used to conduct a step-by-step evaluation that addresses each individual task and the possible influence of the technology on that task. In this way, an understanding of the potential positive and negative impact of the technology can be developed. Possessing a good understanding of the task and how it may change with a new technology is important knowledge that both the customer and the system developer need to have throughout the procurement, design and prototyping stages of developing new technological systems.

Since the various types of task analyses focus on different aspects of a task, e.g. physiological or cognitive demands, there will be a task analysis suitable for evaluating the effects of most elements of most tasks. The earlier a task analysis is conducted in the development process, the more likely it is that problematic features of the design will be identified and fixed before the product reaches a stage where such fixes become more difficult and more costly. A task analysis can assist in ensuring that a new technology or modifications to an existing technology are safe and beneficial. Hence, it is important for both the customer and the system developer to be able to conduct and understand task analyses and the evaluations that are based on them. Although a thorough evaluation of a technology should ideally be conducted by an experienced human factors practitioner, someone without human factors expertise can gather useful information from a task analysis. While it is always best to conduct one's own task analysis or to develop an existing task analysis to one's particular requirements, simply reviewing an existing task analysis will provide the customer with a good understanding of the kinds of questions that need to be asked of the system developer or salesperson. There are many different types of task analysis available and in the next section we review those that are likely to be the most useful in the rail industry.

Different Types of Task Analysis

Hierarchical Task Analysis (HTA)

Hierarchical task analysis was developed in the late 1960s and is the most widely used task analysis method (Kirwan and Ainsworth, 1992). In a hierarchical task analysis, the overall goal is broken down into sub-goals, the operations required to meet those sub-goals, and plans for the order of operations (Kirwan and Ainsworth, 1992). This can be applied to any task from very basic ones such as washing hands, through to complex tasks such as piloting fighter aircraft. Figure 7.1 shows the procedure for conducting a hierarchical task analysis as suggested by Stanton et al. (2005).

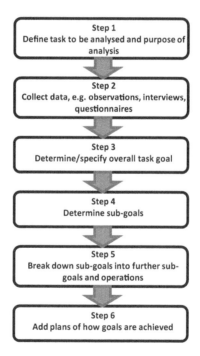

Figure 7.1   Steps in constructing a hierarchical task analysis (adapted from Stanton et al., 2005)

Annett (2005) recommends that, once data has been collected, a draft table or diagram should be prepared and feedback gained from relevant stakeholders. This should then be followed by the identification of significant operations, and the generation and testing of hypothetical solutions to identified performance problems.

Even seemingly basic tasks may have many sub-goals. For example, putting on a pair of trousers may begin with unclipping them from the hanger, opening the waistband, holding open the waist, placing one foot through the first trouser leg, then the other, pulling up the trousers and doing up the zip and button. But this does not cover all the steps. Even placing feet through the trouser legs could have further sub-goals such as leaning forward, bending the knee and raising one leg and possibly leaning against a wall or sitting on a chair or bed. It might also include gathering each trouser leg over the foot and placing the foot back on the floor, and tucking in a shirt. As can be seen from this basic example, a task analysis can be very detailed and for complex tasks, quite lengthy (see Figure 7.2).

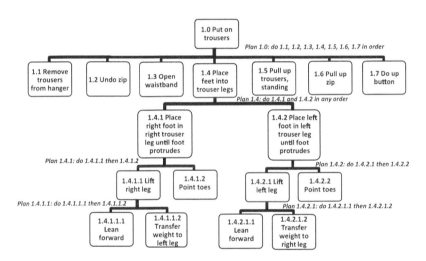

Figure 7.2   Hierarchical task analysis for putting on trousers

Hierarchical task analysis has been used for a variety of purposes including interface design, training and error analysis (Kirwan and Ainsworth, 1992). This method can be performed using pen and paper, requires minimal training to perform and is easily implemented (Stanton et al., 2005). It also provides a comprehensive description of the task under analysis, can be used for a wide variety of tasks in a wide variety of domains, and tasks can be analysed to any level required (Annett, 2005; Stanton et al., 2005). A series of studies conducted by Stanton and Young (1999) found this method to be valid, i.e. the analyses resulting from this method are accurate.

However, hierarchical task analysis has been criticised for not containing any cognitive components, it results in mainly descriptive rather than analytical data, and is not generally useful for design solutions (Stanton et al., 2005). Added to this, the reliability of hierarchical task analysis may be questionable, i.e. analyses of the same task conducted by different analysts may not be the same (Stanton and Young, 1999). Further, it can be time-consuming when analysing large or complex tasks (Annett, 2005; Stanton et al., 2005). Becoming proficient at conducting hierarchical task analyses takes considerable practice and an analyst must be competent in a variety of human factors data collection methods such as observations and interviews (Annett, 2005; Stanton et al., 2005). Finally, the full collaboration of relevant stakeholders is necessary for a successful hierarchical task analysis (Annett, 2005).

Example: Lane, Stanton and Harrison (2006) developed a hierarchical task analysis of the drug administration process used by hospital nurses. The task analysis detailed the many tasks required to administer a drug to a patient with the four top-level tasks being 'Check chart for medication details', 'Acquire medication', 'Administer drug to patient' and 'Record dosage'. Each task then showed the sub-tasks required, sometimes indicating numerous choices for the next step, and sometimes indicating the order in which the tasks should be performed. For example, in order to acquire medication (2.0), the nurse must carry the patient chart to the drug trolley/cupboard (2.1), *or* get the medication from ward stock (2.2), *or* get the medication from the drug trolley (2.3), *or* get the medication from the controlled drug cupboard (2.4) and *then* prepare the medication (2.5). The information derived from the task analysis was then used in a technique designed to predict errors and identify possible solutions aimed at preventing errors or minimising their effects. A section of the task analysis is shown in Figure 7.3.

Goals, Operators, Methods and Selection Rules (GOMS)

GOMS is a method specific to the domain of human-computer interaction. The aim of the process is to define the user's goals, divide these into their relevant sub-goals and describe how user interaction leads to achievement of the user's goals (Stanton et al., 2005). GOMS can be used to describe the way in which a task is performed, to predict times for performing tasks and to predict learning (Stanton et al., 2005). There are four basic components of GOMS: (1) goals, i.e. the overall goal/s and its/their sub-goals; (2) operators, i.e. actions required by the user to achieve the goals, including motor and cognitive actions; (3) methods, i.e. the procedures followed by the user (there can be more than one possible procedure); and (4) selection rules, used to determine which method should be used when more than one method is available (Stanton et al., 2005). As with hierarchical task analysis, the first step of the GOMS process is to define the user's main goals, then decompose those goals into sub-goals. The next step defines the actions required to meet each of those sub-goals, followed by a description of the procedure(s)

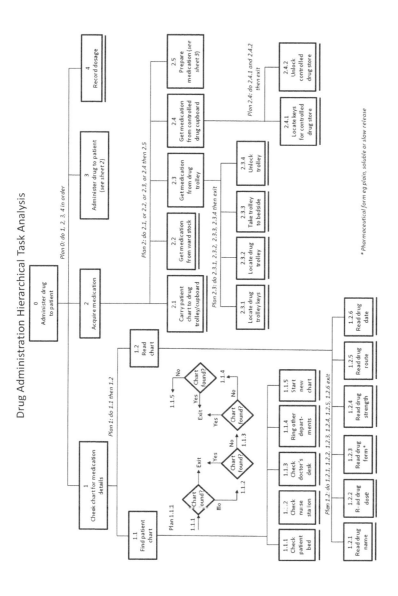

Figure 7.3 HTA for administering drug to hospital patients (Lane, Stanton and Harrison, 2006, reprinted by permission of Elsevier)

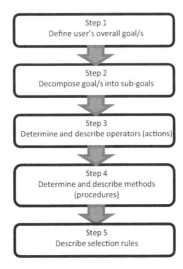

Figure 7.4   Steps in GOMS procedure (adapted from Stanton et al., 2005)

required to meet those goals. Finally, every possible method of achieving the goals should be described and, where more than one method is possible, selection rules should be determined to predict the method that will be used (Kieras, 2004; Stanton et al., 2005). These steps are shown in Figure 7.4 above.

Looking again at our example of putting on a pair of trousers, a GOMS analysis will be similar to the HTA but will include alternative procedures, such as sitting on a chair/bed, leaning against the wall or balancing on one leg without support before putting one leg at a time into the trousers. Selection rules may include the person's ability to balance on one leg (if not capable, sit down or lean against a wall). This is, of course, a very simple example but it demonstrates the requirements of a GOMS analysis.

The main benefit of GOMS is that it provides a hierarchical description of task activity and describes a number of potential routes to performing the same task (Stanton et al., 2005). Performance and learning times may be of benefit for designers in choosing between systems, and GOMS has been applied and validated extensively in the human-computer interaction domain (Stanton et al., 2005). However, it can be a difficult method to apply, is time-consuming, and its use outside the domain of human-computer interaction is limited. Further disadvantages are that GOMS only models error-free, expert performance, and requires a high level of training and practice to become proficient in its use (Stanton et al., 2005).

Example: Beard, Smith and Denelsbeck (1996) describe how they used a simplified GOMS method (QGOMS) to aid the design and ongoing development of an interface for displaying computed tomography medical images. Beard et al. recorded times taken for each step in the process of interpreting CT images using a workstation with current technology and compared them to the times taken for a workstation using new technology. Their first attempt showed that the new technology was slower than the current technology, thus they developed another workstation and conducted further testing. They found that the new workstation was faster than the current one. Figure 7.5 is a copy of their QGOMS analysis for one of the workstations. The numbers shown in the lower section of each box are the times taken to complete those tasks.

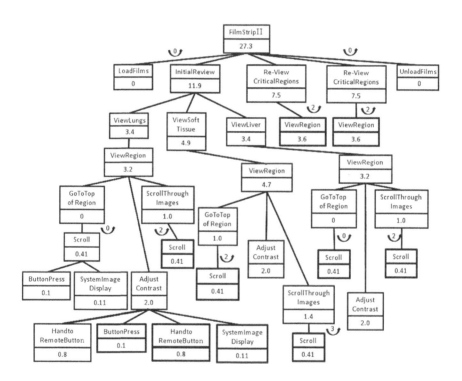

Figure 7.5   QGOMS analysis of procedure for reading computer tomography medical images (Beard, Smith and Denelsbeck, 1996, reprinted by permission of Taylor & Francis, Ltd.)

Cognitive Task Analysis (CTA)

Conventional task analysis methods generally provide a description of the physical activities performed when a person interacts with a complex system. Cognitive task analysis methods attempt to determine the cognitive processes used when performing a task and look at the specific skills and processes employed by users (Stanton et al., 2005). There are several types of cognitive task analysis and we recommend Stanton et al. (2005) for a review of various types.

The general approach to a cognitive task analysis is to conduct observations and interviews of operators and subject matter experts to build an overall picture of the various cognitive elements of a task. As with other task analyses, the overall aim will guide the procedure but it is likely to begin with gathering data (interviews and observations), followed by organising the data into a hierarchy of goals or other predefined parameters. The data may be analysed for various purposes including designing and evaluating a new technology, designing procedures and processes, designing and evaluating training procedures and interventions and evaluating team and individual performance (Stanton et al., 2005).

The obvious advantage of cognitive task analyses is that they can determine and describe the cognitive elements that underlie operators' goals and decision-making (Stanton et al., 2005). Cognitive task analysis methods have been used extensively in numerous domains and have been subjected to numerous validation studies (Stanton et al., 2005). The main disadvantages of this method are that it requires considerable resources in terms of time, effort and participants including subject matter experts, and it requires a high level of skill by the analyst (Stanton et al., 2005).

Example: Seamster, Redding, Cannon, Ryder and Purcell (1993) developed a cognitive task analysis of the cognitive processes used by air traffic controllers to support the development of a curriculum for en route air traffic control. Data were collected via interviews, problem-solving exercises and simulated performance modelling. Seamster et al. identified 13 primary tasks, as follows:

- Maintain situation awareness
- Develop and revise sector control plan
- Resolve aircraft conflict
- Reroute aircraft
- Manage arrivals
- Manage departures
- Manage over-flights
- Receive handoff
- Receive point-out
- Initiate handoff
- Initiate point-out
- Issue advisory
- Issue safety alert.

The data were also used to compile a mental model of an expert air traffic controller's knowledge organisation, to look at strategies used and to highlight a hierarchy of goals. Information derived from this analysis led to recommendations regarding training for the Federal Aviation Authority's revised curriculum for air traffic controllers.

There are many other forms of cognitive task analysis. Below are three variations that are commonly used and potentially the most useful for the rail industry.

Goal-Directed Task Analysis (GDTA)

A goal-directed task analysis focuses on the numerous goals within a given task, the major decisions required to achieve those goals and the situation awareness needs to make decisions (Endsley, Bolté and Jones, 2003; Rose and Bearman, 2012). GDTA does not include the way that information is acquired by the user because of the many different ways that this can be achieved (Endsley et al., 2003). Information acquisition can change because of differences between systems and with advances in technology. Operators may also have their own preferred method of acquiring the information needed for their task and one method may be just as effective as another, thus the method of acquiring information is not considered to be important in a GDTA. In a GDTA the aim is to outline the information that operators require to make decisions that meet each individual goal, regardless of whether or not that information is actually available to the operator (Endsley et al., 2003) (see Figure 7.6).

Figure 7.6   Steps in compiling a goal-directed task analysis (adapted from Endsley, Bolté and Jones, 2003)

The advantage of GDTA is that it provides information that would be helpful in designing systems to support the operator's situation awareness relevant to each goal (Endsley et al., 2003) (see chapter 10 for more information on situation awareness). GDTA has been used to analyse situation awareness requirements in a wide variety of domains, including air traffic control (Endsley and Rodgers, 1994), medicine (Kaber, Segall, Green, Entzian and Junginger, 2006), and manufacturing (Usher and Kaber, 2000). As the elements of a GDTA are based on the information needs of the operator regardless of the technology or system currently in use, the resulting analysis will not be limited to a specific technology or system and would not need to be repeated unless the goals of the operator change (Endsley and Garland, 2000).

According to Endsley et al. (2003), the disadvantage to this method is that it can be costly and time-consuming because compilation commonly takes 3 to 10 sessions with subject matter experts, and it is recommended that 10 to 20 experts review the completed analysis in order to identify any missing information. Obtaining sufficient time with subject matter experts may be difficult (Endsley et al., 2003). The technique also requires a good understanding of situation awareness and proficiency in interviewing skills. For a detailed step-by-step procedure of how to conduct a GDTA, see Endsley et al. (2003).

Example: Usher and Kaber (2000) developed a goal-directed task analysis of the intervening activities required by an operator in a flexible manufacturing system. The overall goal was 'achieve planned output' which had four sub-goals: 'manufacture jobs to meet due dates', 'avoid bottlenecks', 'expedite critical orders' and 'maintain normal system functions'. These were then broken down into further sub-goals. Beneath the sub-goals were the objectives for meeting each sub-goal, along with the sub-objectives, and finally further tasks required to meet those sub-objectives. For example, the sub-goal of 'avoid bottlenecks' had two objectives: 'identify capacity bottlenecks' and 'resolve capacity bottlenecks'. The latter had three sub-objectives: 'keep bottleneck at full capacity', 'reroute jobs to underutilized resources' and 'suspend job(s) with high slack'. Tasks required for the latter sub-objective were 'identify job(s) ahead of schedule' and 'suspend job(s)'. The information derived from this analysis was used to provide guidelines for the presentation of information designed as a decision-making aid for operators. One section of the task analysis is shown in Figure 7.7.

Critical Decision Method (CDM)

The critical decision method is a form of cognitive task analysis that can be useful in investigating the cognitive processes involved in decision-making in non-routine situations or events (Klein, Calderwood and MacGregor, 1989). This method involves selecting a suitable incident, obtaining an unstructured account of the incident, establishing an incident timeline and identifying critical decision points (Klein et al., 1989; Tichon, 2007). The interviewer then probes the participant to draw out more information relating to each critical point, such

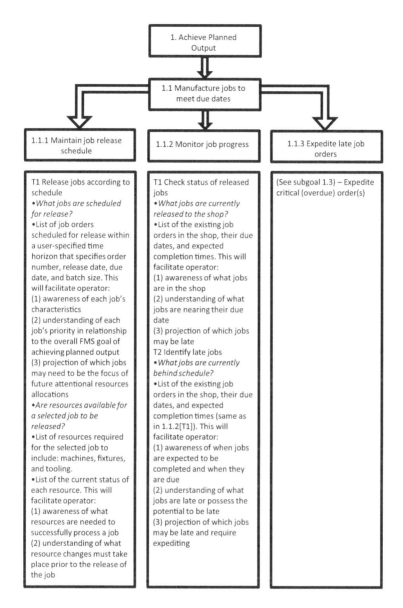

Figure 7.7   One sub-goal of GDTA for achieving Flexible Manufacturing System planned output (adapted from Usher and Kaber, 2000, used by permission of John Wiley & Sons, Inc.)

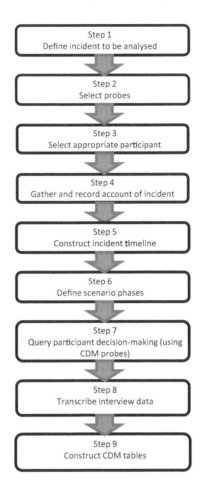

Figure 7.8   Steps in critical decision method (adapted from Stanton et al., 2005)

as goals, strategies and the reasons why participants made the decisions that they did (Klein et al., 1989; Tichon, 2007). Figure 7.8 shows the steps involved in the critical decision method.

There are numerous ways of analysing the information derived from a CDM study and the method chosen will depend upon the ultimate aim of the analysis. For example, a critical cue inventory may be used to gather information about specific cues used in assessing situations and considering choices of action, with the resulting data used to design training that is specific to that situation (Klein et al., 1989). Another form of analysis is to compile a situation assessment record in which critical cues and current goals are probed for each decision point (Klein et al., 1989). This provides a record of situation assessment strategies used by operators and how situation assessment may change due to observed cues that may result in a change to or replacement of previous goals (Klein et al., 1989).

The CDM has many possible applications such as evaluating systems aimed at aiding operators and identifying areas where training could improve performance (Klein et al., 1989). As an example of the latter, a task analysis using the critical decision method could be conducted to highlight differences between expert and novice operators. Training could then be designed to focus on those differences to speed up the process of learning for novice operators and improve their overall performance.

One of the advantages of the CDM is that it provides meaningful information about non-routine events that would not usually be captured in other task analysis methods (Klein et al., 1989). The iterative structure of the critical decision method ensures an in-depth analysis, and the probes and queries have been widely used and found to be fruitful in capturing decision-making processes (Klein and Armstrong, 2005). The CDM is relatively quick and easy to apply (Stanton et al., 2005) and coding of semi-structured interviews is also relatively straightforward (Klein et al., 1989).

A disadvantage of this process is that it relies on verbal reports, which may not be completely accurate (Klein et al., 1989). While the procedures used in the CDM (such as the use of neutral probes, in-depth recall of a situation) have been designed to minimise this potential bias (Klein et al., 1989), the reliance on verbal reports is still a limitation of the method. Another limitation of this method is that it requires a high level of resources for a relatively small dataset, however, the richness of the resulting data may justify the use of resources (Klein and Armstrong, 2005). Using the CDM requires a high level of expertise (Klein and Armstrong, 2005) and the data gathered is highly dependent upon the analyst's skills and the quality of the participants (Stanton et al., 2005). For a detailed step-by-step procedure of how to conduct a CDM task analysis, see Stanton et al. (2005).

Example: O'Hare, Wiggins, Williams and Wong (1998) describe three studies in which they used the critical decision method. The first study looked at the task of white-water rafting and resulted in recommendations regarding training of guides. The second study involved eliciting information from pilots about critical cues for different weather conditions which also led to recommendations for training. In the third study, their aim was to determine display requirements for a computer-based ambulance dispatch system. The CDM probes that O'Hare et al. used are shown in Table 7.1.

Table 7.1   CDM probes used by O'Hare et al. (1998)

Goal specification	What were your specific goals at the various decision-points?
Cue identification	What features were you looking at when you formulated your decision? How did you know that you needed to make the decision? How did you know when to make the decision?
Expectancy	Were you expecting to make this type of decision during the course of the event? Describe how this affected your decision-making process.
Conceptual model	Are there any situations in which your decision would have turned out differently? Describe the nature of these situations and the characteristics that would have changed the outcome of your decision.
Influence of uncertainty	At any stage, were you uncertain about either the reliability or the relevance of the information that you had available? At any stage, were you uncertain about the appropriateness of the decision?
Situation awareness	What information did you have available to you at the time of the decision?
Situation assessment	Did you use all the information available to you when formulating the decision? Was there any additional information that you might have used to assist in the formulation of the decision?
Options	Were there any other alternatives available to you other than the decision that you made? Why were these alternatives considered inappropriate?
Decision blocking – stress	Was there any stage during the decision-making process in which you found it difficult to process and integrate the information available? Describe precisely the nature of the situation.
Basis of choice	Do you think that you could develop a rule, based on your experience, which could assist another person to make the same decision successfully? Do you think that anyone else would be able to use this rule successfully? Why/Why not?
Analogy/generalization	Were you at any time, reminded of previous experiences in which a *similar* decision was made? Were you at any time, reminded of previous experiences in which a *different* decision was made?

(Used by permission of Taylor & Francis, Ltd.)

Applied Cognitive Task Analysis

An applied cognitive task analysis consists of three interview methods aimed at eliciting information regarding the cognitive demands and skills of a task (Militello and Hutton, 2000). Information derived from this method is represented in a format that can be directly translated into applied products such as recommendations for interface improvements or training scenarios (Militello and Hutton, 2000). The three interviews are: the task diagram interview, which provides an in-depth overview of the task; the knowledge audit interview, which highlights the task elements requiring expertise; and the simulation interview, which probes the cognitive processes used by the subject matter expert during the task (Militello and Hutton, 2000). The information gathered from these interviews is then integrated into a cognitive demands table (Militello and Hutton, 2000). See Figure 7.9 for the steps required for this method.

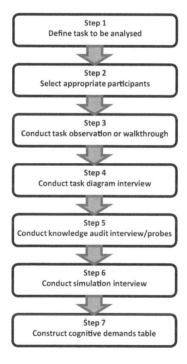

Figure 7.9   Steps in applied cognitive task analysis (adapted from Stanton et al., 2005)

The main advantage of this method is that it is not necessary to have any training in cognitive psychology to be able to use it (Militello and Hutton, 2000). Using three interview methods ensures comprehensiveness, and the provision of probes and questions for the analyst facilitates data extraction (Stanton et al., 2005). Although it is difficult to establish the reliability and validity of cognitive task analyses, Militello and Hutton (2000) do provide some support for the reliability and validity of the method. However, Stanton et al. (2005) state that reliability of the applied cognitive task analysis method is questionable and further validation studies are required. In usability research conducted by Militello and Hutton (2000) participants reported that this method was easy to use and flexible, the outputs of interviews were clear, and the knowledge representations were useful. However, training in the use of this method may be lengthy and, as with many other techniques, the data collection may be time-consuming and data analysis laborious (Stanton et al., 2005). The quality of the data is dependent upon the quality of the experts as well as the skills of the analyst (Stanton et al., 2005). Finally, access to subject matter experts is often difficult (Stanton et al., 2005).

Example: Farrington-Darby, Wilson, Norris and Clarke (2006) incorporated an element of the applied cognitive task analysis method into their naturalistic study of railway controllers. Following interviews and observations of teams and individuals which provided an understanding of the nature of the task environment and the available information in the control room, Farrington-Darby et al. conducted task diagram interviews with subject matter experts. The task diagram interviews were used to elicit details of tasks that are common across events/incidents and types of controllers, and this information was used to compile task diagrams for various events. For example, the task diagram for dealing with a fatality is shown in Figure 7.10.

Figure 7.10   Task diagram for dealing with fatality (Farrington-Darby et al., 2006, reprinted by permission of Taylor & Francis, Ltd.)

This section has outlined a few of the most common methods of conducting a task analysis and some of the strengths and limitations of these methods (see Table 7.2 for a summary). The next section considers different task analyses that have been conducted for various purposes in the rail industry.

Table 7.2   Summary of types of task analysis

Type of Task Analysis	Method	Advantages	Disadvantages
Hierarchical Task Analysis (HTA)	Overall goals, sub-goals, required operations, and plans for order of operations	Useful in many domains. Can be performed using pen and paper. Minimal training required. Easily implemented. Provides comprehensive description of task. Studies showing validity.	No cognitive elements. Laborious and time-consuming for large or complex tasks. Mainly descriptive, not analytical. Reliability uncertain. Requires high level of proficiency in human factors methods.
Goals, Operators, Methods and Selection rules (GOMS)	Goals/sub-goals, operators (actions performed by users), methods (procedures) and selection rules for methods	Provides a hierarchical description of the task. Includes potential routes to performing a task. Includes performance and learning times. Applied extensively in human-computer interaction. Extensive validation in human-computer interaction domain.	Limited use outside of the human-computer interaction domain. Can be difficult to apply. Time-consuming. Only models error-free performance. Requires high level of training and practice to become proficient.
Cognitive Task Analysis	Observations, interviews, consultation with subject matter experts	Can determine and describe cognitive aspects of a task. Has been used widely in numerous domains. Numerous validation studies.	Requires considerable resources, i.e. time, participants, subject matter experts. Requires a high level of skill by the analyst.
Goal-Directed Task Analysis (GDTA)	Overall goal, sub-goals, decisions and situation awareness needs	Provides information to help design systems supporting situation awareness. Used in wide range of domains. Not limited to specific technology or system.	Can be costly and time-consuming. May be difficult to obtain sufficient time with experts. Requires good understanding of situation awareness. Requires proficiency in interviewing skills.

Critical Decision Method (CDM)	Focuses on non-routine situations, establishes a timeline and highlights points where critical decisions are made and looks at goals, strategies and reasons for decisions	Provides meaningful information about non-routine events. Events are real. In-depth analysis. Relatively quick and easy to apply. Coding of interviews is relatively easy.	Relies on verbal reports that may not be accurate. Reliability uncertain. High level of resources relative to dataset. Requires high level of expertise to apply. Data dependent on analyst skills and participant quality.
Applied cognitive task analysis	Combines data from task diagram, knowledge audit and simulation interviews to construct a cognitive demands table	Training in cognitive psychology not required. Comprehensive. Some support for reliability and validity. Flexible and easy to use. Outputs of interviews clear. Useful knowledge representations.	Training may be lengthy. Data collection time-consuming. Data analysis may be laborious. Quality of data dependent on quality of expert and skills of analyst. Access to subject matter experts can be difficult.

Examples of Task Analyses Conducted in the Rail Industry

The best method for developing an understanding of the components of a task is to conduct a task analysis yourself (we outline how to do this for a goal-directed task analysis in 'Constructing a Task Analysis'). However, it is possible to use an existing task analysis and develop it for your own use (Rose and Bearman, 2012). While this is a somewhat contentious point, the basic idea is that conducting a new task analysis every time is unnecessary and that often perfectly good task analyses exist that could be tailored to an organisation's needs. Thus, it is reasonable in many cases to begin with a task analysis that has already been constructed and then adapt it for one's own operation, geographical location and specific purpose. Depending on the task that is being considered, there may be several different task analyses that are available. In such situations it is sensible to choose the one that best fits the purpose. It should be cautioned however that existing task analyses should always be developed further according to one's specific requirements and extensively reviewed by key stakeholders before being used to evaluate a new technology because there is a danger of missing important information that is critical to the task.

To provide a set of basic task analyses, a number of different task analyses published in the literature will be reviewed below. Due to the constraints of space it is not possible to reproduce the full task analysis for each study, however, the references for these papers can be found in the reference list at the end of this chapter. Readers who are interested in using any of the task analyses discussed in this section are directed to these original sources for more information. The following discussion also provides a flavour of the kinds of topics that task analysis has been used to investigate.

Task Analyses Related to Examining New Technology

Examining an In-Cab Advisory Information System

The Rail Safety Standards Board (2009) investigated the potential for an in-cab driver advisory information system to reduce train energy usage and increase network capacity. The advisory information is provided to drivers via a satellite navigation system and includes, where applicable, new target arrival times in situations where conflicts arise due to trains running out of course (Rail Safety Standards Board, 2010). The aim of this device is to enable drivers to employ strategies to minimise their fuel usage in situations where they need to slow down to a slow speed or stop at the approach to a junction or station due to another train blocking their access. According to the Rail Safety Standards Board (2010), providing this information to drivers should assist them in regulating their speed prior to conflict points which should then lead to a reduction in energy consumption, train delays, maintenance costs relating to brake wear and risk of passing signals set at danger (SPADs).

As part of their investigation into this technology, the Rail Safety Standards Board (2009) examined an existing UK Train-Driving Generic Task Analysis, a hierarchical task analysis developed to support an analysis of driver workload. The Rail Safety Standards Board (2009) provide the top level of this hierarchical task analysis with the overall goal of providing a train service, broken down into five stages:

- Preparing for service
- Driving the train
- Performing activities specific to the type of service
- Responding to failures
- Concluding service.

These are broken down into various sub-tasks and plans have been included for the various tasks (Rail Safety Standards Board, 2009). The three sub-goals of 'driving the train' are as follows:

- Start away from scheduled stop (including sub-tasks and plans).
- Drive towards scheduled stop (including sub-tasks and plans).
- Stop for scheduled stop (including sub-tasks and plans).

These sub-goals are broken down further and plans have been included (Rail Safety Standards Board, 2009).

The Rail Safety Standards Board (2009) then developed a cognitive task analysis to describe speed control strategies used by drivers. Their analysis identified:

- information about speed targets used by drivers
- decision-making of the driver regarding appropriate train speeds
- how drivers use train controls to achieve the appropriate speeds.

Decision-making includes:

- collecting and recalling current and future speed targets from route knowledge
- selecting the appropriate train speed
- monitoring the speed of the train
- comparing the appropriate speed with the actual train speed
- using the difference between required speed and actual speed to make decisions about power and brake settings, taking train handling factors such as gradients and curves into consideration (Rail Safety Standards Board, 2009).

Speed control behaviours include cruising, increasing speed, reducing speed, levelling speed (maintaining a consistent lower speed rather than conforming to permitted speed changes) and maintaining the timetable (Rail Safety Standards Board, 2009).

The Rail Safety Standard Board's (2010) findings suggested that considerable cost savings through reduced fuel usage would be possible with the introduction of a driver advisory information system. There are also safety implications with the potential to reduce the risk of SPADs.

Assessing the Effectiveness of an In-Cab Signal Reminder Device
The Rail Safety Standards Board (2005) assessed the potential effectiveness and risks of an in-cab signal reminder device (ICSRD) designed as an addition to the Automatic Warning System (AWS). AWS provides train drivers with an auditory and visual warning of the aspect of approaching signals, as well as warning of temporary and emergency speed restrictions, and some permanent speed restrictions (Office of Rail Regulation, 2008). When the train passes over a magnetic inductor, the visual display will turn black and stay black if the signal aspect is clear and a bell will sound (Office of Rail Regulations, 2008). If the signal aspect is cautionary or danger, the visual display will alternate between black and yellow and a continuous horn will sound until the driver cancels the horn, which must be done within 2 seconds or a full emergency brake application will appear (Office of Rail Regulations, 2008). Following cancellation of the horn,

the visual display continues to alternate between black and yellow as a reminder of the warning (Office of Rail Regulations, 2008). AWS does not distinguish between red, double yellow or single yellow aspects but the ICSRD displays miniature lights that indicate the signal aspect (Rail Safety Standards Board, 2005). By pressing a button, the driver must acknowledge the warning of signal aspect, and the display then remains on as a constant reminder of the aspect of the last signal passed (Rail Safety Standards Board, 2005).

The Rail Safety Standards Board (2005) compared the driving task using standard AWS with the task using the signal reminder device. Their main focus was on correct responses to signal aspects. In combination with a literature review and workshops with subject matter experts, a hierarchical task analysis highlighted several potential advantages and disadvantages of the device. The first part of both analyses (one for using standard AWS and one for using the signal reminder device) describes all tasks associated with driving a passenger train, from entering the cab through to tasks associated with driving along the track, such as monitoring speed, applying brakes and maintaining attention on or near the track. The remainder of both task analyses describes tasks specifically relating to signal aspects and the use of AWS or the signal reminder device. Tasks are described in terms of:

- signal aspects on the route
- AWS warnings received (both AWS and ICSRD warnings are included in the ICSRD task analysis)
- required response from the driver
- consequences of the correct action
- description of some of the issues that arise (Rail Safety Standards Board, 2005).

The task analysis was then used for error analysis, specifically looking at potential errors in using AWS or the signal reminder device, and errors that may arise due to use of the devices. Errors unrelated to AWS use or with no safety consequences were excluded from the error analysis. As an evaluation of the signal reminder device, the error analysis appears to be the most beneficial and useful step in the process. The error analysis not only identified potential errors but also highlighted how errors may be caused, compounded, alleviated, corrected or avoided as a result of using the signal reminder device (Rail Safety Standards Board, 2005).

Developing an Interface to Support Train Drivers' Schedule Management
Tappan, Pitman, Cummings and Miglianico (2011) used a cognitive task analysis to determine the information requirements of train drivers relating specifically to schedule management. The steps taken in compiling the task analysis were firstly to conduct a scenario task overview (similar to a hierarchical task analysis), followed by an event flow diagram which defines when and in what order events and tasks occur. The next step was to compile decision ladders for each complex decision highlighted in the event flow diagram with the aim of understanding the

information required for decision-making. A decision ladder captures an operator's state of knowledge and information processing activities required in decision-making, especially for complex decisions. Steps 4 and 5 involve generating situation awareness and information requirements. The information derived from this analysis was used to aid in the design of an interface that was designed to support drivers' schedule-related decision-making. Seventeen requirements were highlighted:

- Current speed
- Goal speed
- Speed differential
- Current traction/friction lever position
- Current time
- Impact of event on schedule
- Suggested speed profile
- Departure time
- Time to departure
- Potential impact of rail grade on speed
- Upcoming speed change indication
- Scheduling anomaly alert
- Train route with current location
- Next waypoint with scheduled arrival time
- Difference between predicted arrival and scheduled arrival at next waypoint
- New recommended speed profile with impact on schedule
- New recommended goal speed (Tappan et al., 2011).

The resulting display prototype was divided into six functional elements: a summary of high-level trip details, such as city of departure and city of arrival; trip planning information such as delay time; an overview of the route; speed profile bars including maximum speed and economy/suggested speed; a terrain profile; and an information display summarising important data such as predicted arrival at destination, and current and upcoming track gradient (Tappan et al., 2011).

Considering the Effects of New Train Control Technologies on Train Drivers
Roth and Multer (2009) compiled a cognitive task analysis to evaluate the potential impact of new train control technologies on the cognitive demands and activities of locomotive engineers (drivers and training instructors). They conducted interviews and in-cab observations in passenger and freight operations at seven locations in the USA across a five-year period. In the first series of interviews, topics included training and knowledge requirements, start of shift and shift turnover activities, distribution and coordination of responsibilities (when two people were in the cab), cognitive demands, establishing priorities, managing workload, planning and decision-making strategies, radio communications, strategies for maintaining

situation awareness and dealing with fatigue (Roth and Multer, 2009). A second series of interviews were then conducted specifically relating to the potential impact of new train control technologies. Their task analysis of the knowledge and cognitive functions involved in operating a train is summarised in Figure 7.11.

The cognitive task analysis revealed design implications for in-cab displays, including the importance of ensuring that information displayed can be accessed quickly and without diverting attention away from the external environment. The results indicated that some cognitive demands would be reduced, such as trying to keep track of temporary speed restrictions, while new cognitive demands were created, such as monitoring the train control displays inside the cab. Training implications arising from the task analysis were not only the obvious need for training on how to use and interpret the train control system but also how to use the display appropriately without being distracted from events outside the train. Another suggestion was for training to include handling the train without the train control system to ensure drivers are still capable of doing so in the event of failure of the train control system (Roth and Multer, 2009).

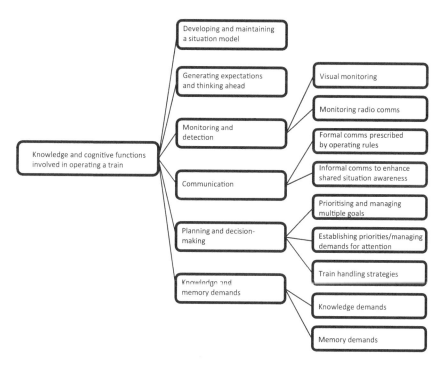

Figure 7.11   Task analysis of knowledge and cognitive functions involved in operating a train (adapted from Roth and Multer, 2009)

A New Communication Device for Train Dispatchers

Roth, Malsch, Multer and Coplen (1999) compiled a cognitive task analysis of railroad dispatching in order to identify the strategies used by dispatchers to control track use. Data for the analysis was collected using observations of dispatchers (an arm of train control) at both passenger and freight operations, and interviews. They found that dispatchers use many strategies to ensure the safe and efficient scheduling of trains, a task made complicated by the need to communicate and coordinate with other dispatchers, drivers and maintenance of way workers. Effective communication is required to deal with unplanned demands such as accommodating unscheduled trains and unanticipated events such as train delays; to keep track of multiple trains, including some that are not in their area of control; and to acquire and maintain knowledge of factors affecting speed such as characteristics of the trains, weather and track conditions (Roth et al., 1999). Two of the strategies described by Roth et al. (1999) involve anticipating and planning ahead, and acting proactively. Anticipating and planning ahead includes maintaining awareness of the bigger picture by not only monitoring activity within their own territory but also the activity of trains in other territories that may impact upon their decision-making (Roth et al., 1999). The necessary information is gathered from:

- displays showing trains within their territory
- notifications from other dispatchers
- radio communications regarding trains in other territories (Roth et al., 1999).

Also part of anticipating and planning ahead is cooperative planning with other dispatchers to facilitate train movement across territories and maximise efficiency (Roth et al., 1999). Dispatchers also use the 'party line' feature of radio communications whereby they listen to radio communications meant for others that may be relevant and helpful in planning for potential delays or problems (Roth et al., 1999). The types of information dispatchers listen for include:

- when a train has departed a station
- equipment problems
- commitments made by others that may affect train movement in their territory (which may assist in avoiding potential hazards)
- mistakes that may arise from confusion or misunderstanding (Roth et al., 1999).

The analysis highlighted the need for a digital communication system to have the ability to broadcast information to all dispatchers in order to avoid losing this important source of information.

Acting proactively involves using strategies that make the most of opportunities such as offering time to a track worker when the track is available (Roth et al., 1999). Another proactive strategy is to increase communication efficiency. For example, a dispatcher may advise a driver approaching a station that there are no messages and so the driver can depart the station as soon as he or she is ready, which allows the driver to head back more quickly (Roth et al., 1999). These data were used to investigate the implications of digital information systems and also led to recommendations for training (Roth et al., 1999).

Other Useful Task Analyses Not Related to Examining a New Technology

Examining Degraded Track Conditions
Tichon (2007) undertook a cognitive task analysis of train driving, using the critical decision method, to inform the development of simulations aimed at improving train drivers' decision-making under degraded track conditions. The critical decision method focuses specifically on decision-making rather than on the overall complex task. Tichon held a focus group discussion with eight experienced Sydney (Australia) suburban train drivers to elicit information about their decision-making methods. Incidents that required a high level of decision-making and which could be replicated in a simulator were selected for further analysis. In the next step of the process, drivers were asked to give a brief description of the selected incidents. An incident timeline was then constructed and decision points were identified. Finally, drivers were probed regarding different aspects of the decision-making process, which included event cues/knowledge, expectations, goals, options, experience and action selection.

The task analysis resulted in seven incidents where drivers are required to make decisions during high workload. The table provided by Tichon (2007, p. 185) lists these incidents, together with the actions expected to be carried out by an expert driver and the errors that could arise if the appropriate action is not taken (see Table 7.3). This type of task analysis can be used for specific purposes, such as designing or evaluating a technology that could impact upon decision-making in critical situations.

Table 7.3   List of incidents, actions and potential errors arising from CDM taken from Tichon (2007)

Incident and cues	Expected action from expert model	Possible errors
Worksite operating under track work authority 2.5km ahead *Cues* Encounter two detonators Outer handsignaller – 'Caution Handsignal'	Apply brakes Slow to 25kph Sound whistle	Does not apply brakes Retains current speed Delay in decreasing speed Does not slow to correct speed Does not sound whistle
Worksite operating under track work authority 500m ahead *Cues* Inner handsignaller – 'Stop handsignal' Encounter three detonators	Sound whistle to acknowledge detonators Sound whistle Stop train	Maintain current speed Delay in slowing train Travelling too fast to stop in time Slow speed instead of stop Stopped train too soon Misinterpret detonators Do not know correct speed Do not know correct procedures Driver frozen – no action
Failed level crossing ahead *Cues* Receipt of warning radio call while driving Receipt of condition affecting network form at station Signal at end of station platform indicates 'medium speed'	Bring train to controllable speed Read form Complete missing fields Depart station at regular track speed Bring train quickly to and travel at regular track speed (approx 80kph)	Does not reduce speed Does not read form Does not understand form Does not complete missing fields Does not respond to speed postage Travels at incorrect track speed
Trespassers on track *Cues* Children crossing track before tunnel entrance	Sound whistle to warn children Slow train speed Report situation to signaller via radio After passing children resume normal track speed Maintain normal speed for 4.5km	No whistle Does not slow train Stops train Does not radio signaller to report incident Makes incorrect report
Condition Affecting Network *Cues* Failed level crossing Handsignallers	Reduce train speed approx 800m before level crossing	Did not slow train Distracted by prior events and forgot failed level crossing

Approaching absolute signal at stop	Depart from station at 'crawl' speed	Stop train to answer radio call
Cues	Maintain slow speed which enables train to stop at failed signal	Slow train to answer radio call
Receives radio call warning of failed signal		Does not slow train after receipt of call
Advised of signal number		Departs station too fast
Directed to call signaller when arrive at failed signal		Travelling too fast, does not correct speed
		Travelling too fast as a result of not compensating for power-up in response to hill
Absolute Signal at Stop	Stop train at signal	Drove past signal
Cues	Radio to seek permission to pass signal quoting correct rule	Misunderstood radio signal
Automatic airbrake activated	Contact guard via intercom to explain situation	Did not wait for permission to continue past
Train stops	Give guard correct bell signal	Failure to use bells at all
	Wait for guard to give correct bell signal back	Use of intercom only to communicate with guard
	Restore brake pipe pressure via gauge	Does not restore brake pipe pressure
	Correct bell to guard	Use of incorrect bell procedure
	Wait for guard to bell back	Does not wait for bell from guard
	Continue at restricted speed to next station	Does not inform guard
		Speed up and travel at too fast track speed

(Used with kind permission from Springer Science and Business Media)

Developing a Unified Model of Train Driver Behaviour

Oppenheim, et al. (2010) are developing a unified model of driver behaviour applicable to all surface transport modes. The model will include the interaction of drivers with novel technologies in emergency situations. The underlying assumption is that factors influencing behaviour are consistent across transport modes. They discuss several factors that can affect safety while driving motor vehicles and trains. These include:

- attitudes and personality, including sensation-seeking
- experience, including hazard perception skills
- driver state (physical and mental ability), including fatigue
- workload
- culture (Oppenheim et al., 2010).

Oppenheim et al. (2010) suggest that the model can be used in the design and safety assessment of new technologies. As part of their model building, they compiled a task analysis of train driving, the higher levels of which are shown in Figure 7.12 below. A detailed analysis of sub-tasks 1 and 2 is provided in their report.

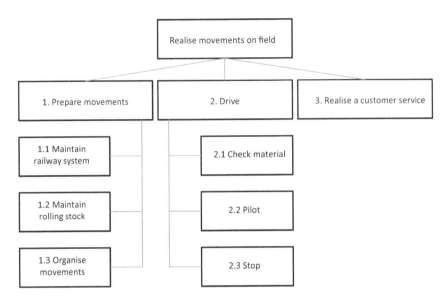

Figure 7.12   **Higher levels of a train-driving task analysis (adapted from Oppenheim et al., 2010, used by permission of the ITERATE Project)**

Constructing a Task Analysis

Although one can use a published task analysis as a basis for developing a task analysis specific to individual needs, this is not always possible as a suitable task analysis may not be available, either because a suitable one has never been done or it is not in the public domain, thus it may be necessary to construct one's own task analysis. To provide an example of how to construct a task analysis, this section describes the process used by Rose and Bearman (2012) to develop a goal-directed task analysis of the long-haul train-driving task. The steps taken are shown below (see also Figure 7.13).

Rose and Bearman's (2012) first step in the process of constructing a goal-directed task analysis was to hold a number of focus group discussions with train drivers to gather information about how the drivers dealt with various aspects of driving such as fatigue, avoiding SPADs, driving in critical zones, maintaining situation awareness, managing radio communications, minimising fuel usage and wear and tear, and any other goals they considered relevant. The discussions were semi-structured, i.e. a few questions were pre-determined but drivers were encouraged to talk about anything they liked and any interesting comments made by drivers were followed up with further questions. The base questions are shown in the box on the following page.

Focus Group Questions:

- What do you find particularly distracting while driving a train?
- What do you think are the most important things to pay attention to when driving a train:
    - when approaching a signal?
    - when in a speed restriction?
    - in other situations?
- Please describe situations where maintaining good situation awareness is a challenge.
    - Please provide some examples of situations where you had low situation awareness.
- What strategies do you use to maintain situation awareness?

It should be noted that conducting focus group discussions with workers must be handled carefully to ensure confidentiality. Workers must be guaranteed that they can speak freely without incurring negative repercussions.

For the second step, the two researchers separately boarded a number of trains to observe the driver and secondary driver (often referred to in Australia as the 'second driver' or 'driver's assistant') while they carried out their usual driving tasks. In Australia, most long-haul freight trains have two drivers working together and although the onus for train handling and communications is on the driver who is currently in control of the train, the second driver's role is to act as a support and ensure that the driver is aware of all relevant information throughout the trip. Both drivers must remain alert at all times. The purpose of having two drivers for long-haul driving is mainly to guard against driver errors that may arise through issues such as driver fatigue, which is a major problem in long-haul train driving. Relay teams are often used in Australia for very long distances, whereby two drivers will be in the cab while two drivers are resting or sleeping and they will change places after a set number of hours. The drivers observed in this step were two-person crews. Drivers were asked to think aloud and comment on what they were doing and why. Whenever a driver carried out an action without talking about it, they were immediately prompted by the researcher, or if the train was in a critical zone (e.g. approaching a yellow and/or red signal), were questioned once they were out of the critical zone.

Whenever it was safe to do so, the researchers asked the drivers and assistant drivers to talk generally about the driving tasks (i.e. not just what they were doing at the time). This proved to be a rich source of data as a great deal of information was derived about events that did not occur during the observational runs. This thinking aloud protocol is useful not only for learning about the how and why of tasks but also the decision-making processes that cannot be observed. The researchers were able to observe drivers' train handling through undulating territory, crossovers, temporary speed restrictions, crossings, entering and leaving stations and, on one occasion, a penalty brake application. Added to this, the researchers could observe how drivers handled radio communications and used cross-calling (calling signals to each other as a way of maintaining situation awareness). For more information about observational data collection, see chapter 5 on qualitative methods.

For the next step, the information gathered during the focus group discussions and the observations was sorted and compiled into specific goals. For example, actions aimed at using the gradient of the track to make decisions on braking and accelerating would be placed under 'controlling speed'. In some cases, actions may have more than one purpose and so would be placed under more than one heading. For example, actions aimed at using the gradient of the track to minimise braking would be placed under 'minimising wear and tear' as well as 'driving in a safe manner'. From this information the overall goals of train driving were defined, then the various sub-goals relating to each major goal were identified. These details were then placed in a hierarchy of tasks, detailing all major and sub-goals required to achieve the overall goal of driving the train from its origin to its destination safely, efficiently and in a timely manner.

Next, the task analysis was shown to a subject matter expert who had extensive experience in the rail industry, including driving and driver training. That subject matter expert made recommendations for improvements to the task analysis and those changes were subsequently made. The modified task analysis was then shown to two other subject matter experts who suggested a few additional changes. Following further consultation with the first subject matter expert, these additional changes were made and the resultant task analysis was finalised. The completed task analysis was used to evaluate a new technology designed to provide information to train drivers.

The example we have provided in Figure 7.13 shows how to construct a goal-directed task analysis that would be useful when the analyst is interested in the situation awareness needs of operators. However, this type of task analysis would not be appropriate for all purposes and tasks. It is important to choose the right type of task analysis for your specific purpose. In the above example, the requirement for the task analysis involved a cognitive aspect (situation awareness) so one of the cognitive task analysis methods was chosen (GDTA). However, if you wish to gather information with the aim of redesigning a train cab with a view to optimising location of controls, a hierarchical task analysis that looks at the physiological tasks (as opposed to cognitive tasks) may be a reasonable choice. If the focus is on specific types of incidents or your interest lies in how operators deal with specific situations, the critical decision method may be the most suitable method. When deciding which method might be suitable, it is often a good idea to consider the literature on task analysis and to consult key stakeholders and other knowledgeable people (both inside and outside the organisation).

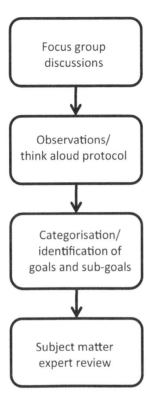

Figure 7.13 Steps for constructing a goal-directed task analysis (adapted from Rose and Bearman, 2012)

Example of Applying Task Analysis to Evaluate a New Technology

The following describes a step-by-step evaluation of a new technology using a goal-directed task analysis. The goal-directed task analysis is particularly useful for examining in-cab information devices since it considers situation awareness, which is something that many in-cab information devices purport to improve. The task analysis used for this evaluation can be found in Rose and Bearman (2012). The technology evaluated here is Freightmiser (see Figure 7.14) which is an in-cab information advice system aimed at saving fuel and assisting on-time running (Howlett, Pudney and Vu, 2008). In order to follow this step-by-step evaluation, it will be necessary to view the Rose and Bearman (2012) task analysis and the following details of Freightmiser together. For the purpose of the analysis, it is assumed that the context is a large, long and heavy-haul operation where there are two drivers in the train cab at all times.

The Freightmiser (Howlett et al., 2008) displays information on a small computer screen situated in the train cab. Information displayed is as follows:

- Route information below and behind the train (total 2 kilometres or 1.24 miles) and for 6 kilometres (3.73 miles) ahead of the train:
  - Track elevation
  - Track curvature
  - Trackside features (signals, crossing loops, kilometre posts and level crossings)
  - The location of the train relative to the track and trackside features
- Ideal speed profile displayed in colour coding: red for braking; green for full power; white for coasting
- Track speed limits, including temporary speed restrictions (updated daily), and current train speed
- Destination name. A driver can select alternative destinations.
- Estimated time of arrival at destination shown
- Coloured bars indicating which of seven speed profiles are being followed, ranging from least to most efficient. A driver can select any of the seven profiles depending on their requirements, i.e. a faster and less efficient profile resulting in an earlier arrival time, or a slower and more efficient profile resulting in a later arrival time (Howlett et al., 2008).

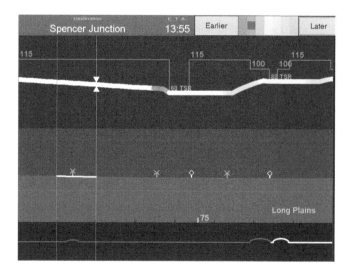

Figure 7.14   Screenshot of Freightmiser (Howlett, Pudney and Vu, 2008, reprinted by permission of the University of South Australia)

There are two possible approaches to evaluating a new technology using a task analysis. The first is to individually examine each feature of the technology and cross-check with each goal and sub-goal of the task analysis to see if there is potential for that feature to impact on each of those goals or sub-goals. The alternative is to examine each goal and sub-goal and look at each element of the technology to look for potential impact. Either of these two approaches should result in a thorough and detailed evaluation, and one approach should be just as beneficial as the other. We have chosen to use the latter approach below, i.e. beginning with the task analysis. (Refer to the task analysis by Rose and Bearman, 2012, for the following evaluation.)

The aim of Freightmiser is to reduce fuel usage and assist in on-time running, which are goals 3.0 and 4.0 of the task analysis. This suggests that Freightmiser may be beneficial but deeper evaluation is necessary to ensure that this is the case, and also to see if there are any other potential benefits or hazards. Some of these hazards are discussed in chapter 2 and some examples are given in the following evaluation of Freightmiser.

The first goal in the task analysis is to execute driving in a safe manner (1.0). The first sub-goal is to monitor and respond to external events (1.1), which requires compliance with relevant signals (1.1.2), complying with relevant speed limits (1.1.3) and negotiating level crossings (1.1.4). These sub-goals then require looking for signal aspects (1.1.2.1), beginning to slow at a yellow signal (1.1.2.2), stopping before a red signal (1.1.2.3), looking for speed boards/signals including temporary speed restrictions (1.1.3.1), beginning to slow in a timely manner (1.1.3.3) and looking for level crossings (1.1.4.1). The information provided on Freightmiser includes location of trackside features including signals and level crossings, as well as the current speed limit and speed limits for six kilometres ahead of the train, including temporary speed restrictions, thus it has the potential to be beneficial in meeting all these sub-goals. Also included under the goal of driving in a safe manner (1.0) is driving to avoid high levels of buff and draft (1.4) and driving to avoid excessive use of brakes (1.5), which are detailed in sections 2.1 and 2.2 of the task analysis and are therefore covered below.

The second goal in the task analysis is to execute driving to minimise wear and tear (2.0) which includes driving to avoid high levels of buff and draft (2.1). In order to meet this sub-goal, it is necessary for the driver to plan strategies to minimise buff and draft (2.1.1.) which requires visualising the track profile below the train and for up to five kilometres ahead of the train (2.1.1.1). It is also necessary to regulate power and brake applicable to the track profile (2.1.2). The gradient and curvature information provided by Freightmiser is likely to be of assistance in meeting these sub-goals.

Driving to avoid excessive use of brakes (2.2) is also required to meet the goal of minimising wear and tear (2.0). This requires planning strategies to minimise use of friction brake (2.2.1), optimising use of kinetic energy (2.2.2) and beginning to slow well before speed restrictions/red signals to avoid heavy use of brakes (2.2.3). All of these sub-goals require visualisation of the track profile below the train and for up to 5 kilometres ahead of the train (2.2.1.1.1, 2.2.2.1 and 2.2.3.1). With the track information display, including gradient, curvature and trackside features, Freightmiser could potentially assist in meeting all of these sub-goals. Also, to optimise use of kinetic energy (2.2.2), it is necessary to drive at speeds to minimise braking while maintaining on-time running (2.2.2.2). With the ability to select from seven potential profiles with varying degrees of speed and efficiency, together with the ideal speed profile indicating when to coast, when to apply full power and when to brake, Freightmiser could be beneficial in achieving these sub-goals.

Executing driving to maintain on-time running (3.0) requires regulation of power and brake applicable to the track profile (3.1), which requires visualisation of the track profile below the train and for up to 5 kilometres ahead of the train (3.1.1). It is also necessary to target speed restriction locations to ensure optimal speed is achieved and maintained (3.2), which requires accurately predicting deceleration and acceleration rates applicable to the train and terrain (3.2.1), and accurately predicting brake application and release lag times applicable to the train and terrain (3.2.2), both of which require visualisation of the track profile below the train and for up to 5 kilometres ahead of the train (3.2.1.1 and 3.2.2.1). Clearly, the track information provided by Freightmiser has the potential to assist in meeting these sub-goals. The trackside features display would be of assistance in meeting sub-goal 3.2 in particular.

Executing driving to optimise fuel usage (4.0) requires regulating power and optimising dynamic brake usage applicable to the track profile (4.1). This requires accurately predicting deceleration and acceleration rates applicable to the train and terrain to avoid slowing more than necessary (4.1.1), and accurately predicting brake application and release lag time applicable to train and terrain to avoid slowing more than necessary (4.1.2), both of which require visualisation of the track profile below the train and for up to 5 kilometres ahead of the train (4.1.1.1 and 4.1.2.1). Once again, the route information displayed on Freightmiser is likely to assist in meeting these sub-goals.

Another sub-goal of 4.0 is optimising use of kinetic energy to avoid excessive use of the throttle (4.2), which requires monitoring of speed (4.2.1), monitoring location for track gradient (4.2.2) and planning strategies to optimise use of kinetic energy (4.2.4) with the sub-goal of visualising the track profile below the train and for up to 5 kilometres ahead of the train (4.2.4.1). Monitoring speed may be assisted by the display of the track speed limit on Freightmiser, while visualising the track profile could again be assisted by the route information, including track elevation and curvature.

This evaluation suggests that Freightmiser has the potential to minimise fuel consumption and assist in on-time running, as it was designed to do. However this evaluation also highlights the potential for Freightmiser to assist with driving in a safe manner with the provision of information on trackside features, current and future speed limits and temporary speed restrictions, as well as minimising wear and tear on couplings (by minimising buff and draft) and brakes. Other goals

(5.0 fix physical problems with train and track, and 6.0 attend to personal needs) shown in the task analysis have not been mentioned in the above evaluation as Freightmiser is not likely to have a direct impact on those goals.

The analysis identified a number of hazards with the use of Freightmiser to meet the higher-level goals (1.0, 2.0, 3.0, 4.0) discussed above. Perhaps the main hazard is the potential for distraction caused by the introduction of an interface in the train cab. Distraction caused by the interface may result in a reduction in the ability of the driver to meet any of the goals and sub-goals, but of particular concern would be a reduction in meeting any of the sub-goals of Goal 1.0 'Driving in a Safe Manner'. This problem may be overcome or alleviated to some extent with appropriate training such as familiarisation training. The relevance of the information on the display can also be examined with regard to meeting goals to ensure that all the information provided is relevant and useful. This is important to prevent degradation of situation awareness due to an excess of irrelevant information (Roy, Breton and Rousseau, 2007). The analysis of Freightmiser presented above suggests that most of the information displayed appears to be relevant to the train-driving task. The only element not mentioned in the evaluation is the estimated time of arrival, which is necessary for selecting the appropriate speed profile, thus all elements appear to be useful. There may be the potential for drivers to become dependent upon the interface and thereby lose vital skills that may be needed should the interface be unavailable for any reason. This potential hazard may be overcome or alleviated with regular simulator training without the use of Freightmiser.

Task analysis is a vital step in the process of evaluating a new technology. As mentioned at the beginning of this section, an evaluation based on a task analysis will not provide a definitive assessment of the effectiveness or potential detrimental effects of a technology. However, it provides a useful way of thinking through the potential costs and benefits that may occur and to potentially highlight a wide range of human factors issues. These issues can be used to shape the design process and highlight areas that may need to be investigated further using different, more focused research methods (e.g. using mock-ups or situation awareness measures). It should also be noted that on paper it may appear that a new technology is going to be useful and worthwhile, but problems may arise during or after implementation. It is important therefore to continually assess the impact of a new technology, using the variety of methods outlined in the different chapters of this book. The technology should also be evaluated in terms of the other technologies that already exist in the train cab or train control room and the social context into which the technology is being placed. Even if a new technology is found to be beneficial and free of potentially serious negative consequences, implementation of that technology may not necessarily be widely accepted by the operators. Resistance to technology is a very common phenomenon and should be taken into consideration when planning the introduction of a new technology. Chapter 3 discusses the issues around resistance to technology in more detail.

Conclusion

Task analysis is a flexible way of providing information that can inform the development and evaluation of a new technology. Task analysis can be used in different ways. At the lowest level, it can be used as the basis for developing an understanding of potential issues that need to be addressed by the developer of a technology. At a more sophisticated level, task analysis can be used to conduct a thorough evaluation of the likely impact of a new technology on tasks performed by the operators. This chapter has sought to explain what a task analysis is, how it has been used in the rail industry and how it can be used in the design and evaluation of new technologies. We have examined the different methods of conducting a task analysis and provided some guidelines on choosing the correct method. Finally we outlined a step-by step process for conducting a task analysis and showed how task analysis can be used to evaluate an in-cab information device. Because of its flexibility and the basic level knowledge that it provides, task analysis is a key method that allows organisations to inform the development and evaluation of new rail technologies. This ensures that new technologies do not possess human factors issues that may be inherently unsafe and extremely costly to resolve.

Acknowledgements

The authors are grateful to the CRC for Rail Innovation (established and supported under the Australian Government's Cooperative Research Centres program) for the funding of this research. We are also grateful to the subject matter experts for their assistance in compiling the task analysis for the long-haul train-driving task, and to the drivers for participating in the focus groups and the observational study. A special thanks to Robert Noy for his assistance in organising the focus group discussions, observations and task analysis reviewing, as well as answering a multitude of questions put to him by the authors. Thanks also to Dr Vincent Rose for his assistance in redrawing many of the figures in this chapter.

References

Annett, J. (2003). Hierarchical task analyses. In E. Hollnagel (ed.), *Handbook of Cognitive Task Design* (pp. 17–35). Mahwah, NJ: Lawrence Erlbaum Associates.

Annett, J. (2005). Hierarchical task analysis (HTA). In N. Stanton, A. Hedge, K. Brookhuis, E. Salas and H. Hendrick (eds.), *Handbook of Human Factors and Ergonomics Methods* (pp. 33/1–33/7). Boca Raton, FL: CRC Press.

Annett, J., Duncan, K.D., Stammers, R.B., and Gray, M.J. (1971). *Task analysis* (Training Information Paper No. 6). London: Her Majesty's Stationery Office.

Beard, D.V., Smith, D.K., and Denelsbeck, K.M. (1996). Quick and dirty GOMS: a case study of computed tomography interpretation. *Human-Computer Interaction, 11*, 157–80.

Crystal, A., and Ellington, B. (2004, August). Task analysis and human-computer interaction: approaches, techniques, and levels of analysis. *Proceedings of the Tenth Americas Conference on Information Systems* (Paper 391). New York: Association for Information Systems.

Endsley, M.R., Bolté, B., and Jones, D.G. (2003). *Designing for Situation Awareness: An Approach to User-Centered Design*. London: Taylor & Francis.

Endsley, M.R., and Garland, D.J. (2000). *Situation Awareness Analysis and Measurement*. Mahwah, NJ: Lawrence Erlbaum Associates.

Endsley, M.R., and Rodgers, M.D. (1994). *Situation Awareness Information Requirements for En Route Air Traffic Control* (Technical Report DOT/FAA/AM-94/27). Washington, DC: Office of Aviation Medicine, United States Department of Transportation, Federal Aviation Administration.

Farrington-Darby, T., Wilson, J.R., Norris, B.J., and Clarke, T. (2006). A naturalistic study of railway controllers. *Ergonomics, 49*, 1370–94.

Howlett, P., Pudney, P., and Vu, X. (2008). *Freightmiser: An Energy-Efficient Application of the Train Control Problem*. Retrieved from http://eng.monash. edu.au/civil/assets/document/research/centres/its/caitr-home/prevcaitr proceedings/caitr2008/vu-et-al-caitr2008.pdf.

Kaber, D.B., Segall, N., Green, R.S., Entzian, K., and Junginger, S. (2006). Using multiple cognitive task analysis methods for supervisory control interface design in high-throughput biological screening processes. *Cognition, Technology and Work, 8*, 237–52.

Kieras, D.E. (2004). GOMS models and task analysis. In D. Diaper and N.A. Stanton (eds.), *The Handbook of Task Analysis for Human-Computer Interaction* (pp. 83–116). Mahwah, NJ: Lawrence Erlbaum Associates.

Kirwan, B., and Ainsworth, L.K. (1992). *A Guide to Task Analysis*. London: Taylor & Francis.

Klein, G., and Armstrong, A.A. (2005). Critical decision method. In N. Stanton, A. Hedge, K. Brookhuis, E. Salas and H. Hendrick (eds.), *Handbook of Human Factors and Ergonomics Methods* (pp. 35/1 35/8). Boca Raton, FL: CRC Press.

Klein, G.A., Calderwood, R., and MacGregor, D. (1989). Critical decision method for eliciting knowledge. *IEEE Transactions on Systems, Man, and Cybernetics, 19*, 462–72.

Lane, R., Stanton, M.A., and Harrison, D. (2006). Applying hierarchical task analysis to medication administration errors. *Applied Ergonomics, 37*, 669–79.

Liljegren, E. (2006). Usability in a medical technology context assessment of methods for usability evaluation of medical equipment. *International Journal of Industrial Ergonomics, 36*, 345–52.

Militello, L.G., and Hutton, R.J.B. (2000). Applied cognitive task analysis (ACTA): a practitioner's toolkit for understanding cognitive task demands. In J. Annett and N.A. Stanton (eds.), *Task Analysis* (pp. 90–113). London: Taylor & Francis.

Office of Rail Regulation. (2008). *Automatic Warning System (AWS)*. Retrieved from http://www.rail-reg.gov.uk/server/show/nav.1558.

O'Hare, D., Wiggins, M., Williams, A., and Wong, W. (1998). Cognitive task analyses for decision centred design and training. *Ergonomics, 41*, 1698–1718.

Oppenheim, I., Shinar, D., Enjalbert, S., Dahyot, R., Pichon, M., Ouedraogo, A., and Cacciabue, C., (2010). *Critical state of the art and unified models of driver behaviour*. European Commission DG Research, Iterate, Deliverable No. 1.2, Retrieved from Iterate website: http://www.iterate-project.eu/deliverables/ITERATE_D1.2_v04_First_EC_submission_040210.pdf.

Rail Safety Standards Board. (2005). Assessing the effectiveness of an in-cab signal reminder device. (Report No. MWHA 2004/011, RSSB Ref: T344, 14 October 2005.) London: Rail Safety and Standards Board.

Rail Safety Standards Board. (2009). Driver advisory information for energy management and regulation: Stage 1 report. London: Rail Safety and Standards Board.

Rail Safety Standards Board. (2010). Driver advisory information for energy management and regulation. (Ref: T724, February 2010.) London: Rail Safety and Standards Board.

Rose, J.A., and Bearman, C. (2012). Making effective use of task analysis to identify human factors issues in new rail technology. *Applied Ergonomics, 43*, 614–24.

Roth, E., and Multer, J. (2009). *Technology Implications of a Cognitive Task Analysis for Locomotive Engineers*. Report No. DPT/FRA/ORD-09/03. Washington DC: Federal Railroad Administration: Office of Research and Development,

Roth, E.M., Malsch, N., Multer, J., and Coplen, M. (1999, Sep/Oct). Understanding How Railroad dispatchers Manage and Control Trains: A Cognitive Task Analysis of a Distributed Team Planning Task. *Proceedings of the 43rd Annual Meeting of the Human Factors and Ergonomics Society, 43*(2), 218–22.

Roy, J., Breton, R., and Rousseau, R. (2007). Situation awareness and analysis models. In E. Bossé, J. Roy and S. Wark (eds.), *Concepts, Models, and Tools for Information Fusion* (pp. 27–67). Boston: Artech House.

Seamster, T.L., Redding, R.E., Cannon, J.R., Ryder, J.M., and Purcell, J.A. (1993). Cognitive task analysis of expertise in air traffic control. *The International Journal of Aviation Psychology, 3*, 257–83.

Stanton, N.A. (2006). Hierachical task analysis: Developments, applications and extensions. *Applied Ergonomics, 37*, 55–79.

Stanton, N.A., Salmon, P.M., Walker, G.H., Baber, C., and Jenkins, D.P. (2005). *Human Factors Methods: A Practical Guide for Engineering and Design*. Aldershot: Ashgate Publishing.

Stanton, N.A., and Young, M.S. (1999). What price ergonomics? *Nature, 399*, 197–8.

Tappan, J.M., Pitman, D.J., Cummings, M.L., and Miglianico, D. (2011, July). Display Requirements for an Interactive Rail Scheduling Display. In D. Harris (ed.), *Proceedings of the 9th International Conference on Engineering Psychology and Cognitive Ergonomics* (pp. 352–61). Berlin, Germany: Springer-Verlag.

Taylor, F.W. (1911). *The Principles of Scientific Management*. The Project Gutenberg EBook #6435. Retrieved from https://archive.org/stream/theprinciplesofs06435gut/pscmg10.txt.

Tichon, J.G. (2007). The use of expert knowledge in the development of simulations for train driver training. *Cognition, Technology & Work, 9*, 177–87.

Usher, J.M., and Kaber, D.B. (2000). Establishing information requirements for supervisory controllers in a flexible manufacturing system using GTA. *Human Factors & Ergonomics in Manufacturing, 10*, 431–52.

Chapter 8

Evaluating Your Train Simulator Part I: The Physical Environment

Anjum Naweed

Central Queensland University, Appleton Institute, Adelaide, Australia

Ganesh Balakrishnan

Central Queensland University, Appleton Institute, Adelaide, Australia

Jillian Dorrian

School of Psychology, Social Work and Social Policy, University of South Australia, Adelaide, Australia

Introduction

This book has thus far covered a broad range of important issues relevant to the pursuit of human factors issues in the rail industry. The aim of this chapter is to stray a little closer towards the region straddling the academic–industry plane, and explore the topic of train simulation. Train simulation has rapidly gained momentum in recent years, and in doing so, cultivated a number of different approaches and definitions in the treatment and practice of simulating reality. This chapter contains information that will be very useful for academic researchers, but is primarily aimed at rail industry subject matter experts, particularly senior train drivers, train crew training managers and general managers of train service delivery who are interested in evaluating train simulator technology. It will discuss the most pertinent issues, and clarify simulator design, use and management, towards defining classes of simulator that may be used for various goals.

For general managers in the rail industry, train simulation is a divisive topic; there are those who are convinced by its potential and will procure as many simulators as they can, and there are those who may buy them but perceive them to be little more than cartoons of the driving environment. Nevertheless, those who actually conduct research with train simulators or manage them are best placed to evaluate their potential, gauge trainee response firsthand and ultimately determine the utility of this equipment. Unfortunately, the reality of the matter is that too many train simulators are doing little more than gathering dust. Why? Is it a failure in the technology, or an understanding of the technology? Is there misinformation about different simulator classes, or is it more of an attitudinal issue? The simple answer is that it may be a combination of all of these. This chapter will educate train simulator users by peeling back a few layers to tackle the notion of realism and revisit simulator design issues in order to describe the essence of train driving on railways.

The present and subsequent chapter go hand in hand. For those already in possession of train simulators, this chapter aims to promote rejuvenation into a journey of rediscovery. For those looking to procure a simulator for the first time, this chapter aims to instil the makings of an informed decision. The goal of the subsequent chapter is to describe how train simulators can be used to evaluate performance and technologies, so that industry and academic users, new and old, may realise their full potential and discover the intricacies in human train-driving performance. At its heart, a simulator is an integrated performance monitoring system made up of a physical environment and a psychological or task environment, but in order to evaluate simulators, or use them as an evaluation tool, we need a good idea of the factors that go into designing a simulator, in particular:

- how realism (or fidelity) is defined and interpreted to address the various physical/environmental and psychological/behavioural requirements of simulation
- the level of fidelity that is required to substantially immerse and engage the user, particularly with respect to individual needs of the end user
- how train simulators are discussed and talked about in the rail industry and how the naming convention (nomenclature) impacts a meaningful classification system
- the factors influencing your decisions to specify the physical framework and level of structural realism
- some substantive considerations for industry and research users.

These areas will be considered in turn throughout this chapter. A visual representation of the overall integrated simulator system is displayed in Figure 8.1.

The physical environment frames the simulation and subsequently defines its features and characteristics. These constituents inform the task environment, the nucleus of the integrated working simulator, which has inputs and outputs. The task environment caters for various train-driving tasks, which in turn recruit specific cognitive demands and shape performance measures, though these are all contextualised by different industry and research needs. Most importantly however, the physical and task environments are wrapped up in the notion of fidelity, and the tension between the physical/environmental versus the psychological/behavioural proponents of realism.

There is a strong argument for simulating train-driving operations, but there is also some contention around synthesising realism. Given that simulators aim to replicate realism on many levels, it is also important to question exactly how much realism is enough, and for new technologies and cab layouts, consider the issues around skills transfer, in a bid to understand the peculiarities of rail that confound simulation. These issues will be discussed in 'Central Concepts Underlying

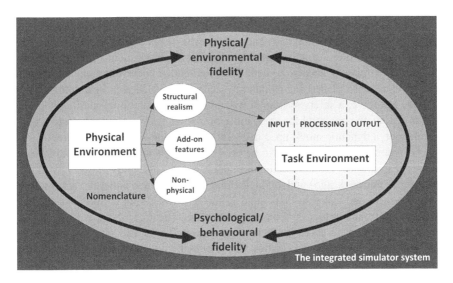

Figure 8.1 **The integrated simulator system showing key elements of the physical environment and connectivity with the task environment**

Simulation'. Indeed, there is a plethora of idiosyncrasies and peculiarities in train driving which render it complex and inimitable in any other system. The rail system and the specific information needs of the task must be translated into the simulation appropriately. This will be illustrated in 'Simulating Train-Driving Operations'.

It is important to note that train simulation has, as a field, been extensively applied in the rail industry, and whilst this has advanced technology, it has also proliferated a counterintuitive naming convention that has gone unregulated, serving only to confound procurement. For that reason, it is useful to classify simulators based on the possible range. The 'Train Simulators' section will identify the features of various train simulators and direct an argument for reconstructing a universal train simulator classification system. Lastly, 'Moving from the Physical Environment to the Task Environment' will explore some important considerations for simulator usage in industry and research, and touch upon material to be discussed at length in the subsequent sister chapter ('Evaluating Your Train Simulator Part 2: The Task Environment'), as the focus shifts from the physical environment to the task itself.

What Can We Evaluate with a Simulator

A train simulator has one purpose, to engage the user in a task that simulates one or more parts of the cab environment, as it is found on the railways. The obvious application for simulators is training – in fact, that is perhaps the main motivation for using a train simulator in industry. Simulators mediate skills transfer and so they are ideally placed for delivering diagnostic skills and procedures, and for providing team training and a dynamic means for assessment (Stanton, 1996). However, simulators can also be used for investigation and evaluation. Simulators are an excellent test-bed for evaluating technology, such as a new information display initiative, a safety system, a warning signal, an auditory alert and so on, but in reality, the potential for evaluation transcends technology through to a new cab layout, a dynamic mock-up, a different seat or an operating procedure.

Of course, simulators can also be used to research substantive human factors issues (e.g. fatigue, sustained attention, decision-making, body support, working postures, climate design) and evaluate human performance under the assumption that it will be the same as that found in the real world, but as the next chapter will go on to show, this is not always the case. The fact is that a simulator *is* a technology and deserves its own evaluation, and it is also important for *it* to be evaluated, otherwise it becomes very difficult to account for the user's own experience of the simulation, and by extension, their acceptance of the simulator.

In summary, both this chapter and the next aim to pave the way for the reader, the industry professional and the academic, to learn a bit more about a simulator's make-up, in order that they may harness it effectively to evaluate a new piece of technology, such as an information display, or conduct empirical research. We also aim for the reader to build a good understanding of how a simulator may be used to evaluate and measure performance for training skills transfer to, say, a new cab environment. Ultimately, a simulator *is* technology, and our aim is that the reader develops a sound understanding of how they may evaluate it to learn about its strengths and limitations and its utility for human performance evaluation.

Central Concepts Underlying Simulation

Simulators in the Rail Industry

Simulators of operational environments are generally designed to replicate the reality, or some part of the reality, found in the live environment. Simulators have, in recent years, become a common tool in the collision avoidance sectors of aviation, maritime and ground transportation (Gray, 2002). Needless to say, a key attraction for using simulation is the ease with which degraded conditions, safety critical scenarios and failure modes can be represented, and the fact that they minimise the need to take live equipment out of service, allowing hazardous situations to be recreated without the real world consequences. Train simulation research generally supports the perspectives underpinning applied adult learning, and numerous studies have shown the benefits of using simulation (e.g. Fukazawa et al., 2004; Keon, 1990), such that simulator-based learning is generally assumed to afford high consistency, availability and throughput, translating to better knowledge and skills retention.

Simulators may also be used for trialling and evaluating technology, and testing new processes and procedures. In the UK, interest in train simulation was originally piqued by the recommendations of the Ladbroke Grove and Southall joint inquiries (Cullen, 2001), and the guidance notes from the Rail Safety and Standards Board (2007b), which impacted the rail industry on an international stage. In Australia, simulator procurement programmes are increasingly rolled into franchise bids and the development of new traction. A decade or so later, one might argue that rail operators are in a position to evaluate their technology more thoroughly and assess the returns on hardware investments. But first, how faithfully do simulators replicate the environment, and how does one define its realism? An issue central to this is the concept of fidelity.

> The end user of a simulator experiences the *gestalt*, which means that all of the parts add up to form an overall impression, even though they may be chosen individually. This means that the whole must be considered, not just the parts.

The Concept of Fidelity

Much as in the matrimonial sense, simulator fidelity describes the faithfulness and loyalty that resides between the simulator and the environment being simulated. Simulator fidelity is a contentious issue, arguably *the* most contentious issue in human reliability research and practice. When considering simulator purchase, there are many fidelity-related questions that are often considered. Is it enough that the simulator looks like a real train? How important is it to have all the same physical aspects (i.e. cab, seat, controls)? How important is the visual representation of the world outside the cab? Does the quality of computer-generated graphics matter? Is it better to pick a real environment and try to recreate it exactly, or to choose a generic platform in which the visual representation of the outside world can be manipulated? Why do I need high fidelity? What is the relationship between fidelity and how seriously people will engage with the simulator? Is higher fidelity always better? What factors encourage people to become immersed in a simulation? Do I need to simulate movement? Questions abound, and turning to the literature does not provide simple, clear answers to these issues. In fact, experts have not even been able to agree on a solid foundation from which to address such questions – a basic definition of fidelity.

Lane and Alluisi (1992) reviewed the topic quite comprehensively but found no single definition for fidelity, and attempts to expound it produced at least 22 separate definitions popularly associated with its key dimensions. These included the psychological, perceptual, functional, physical and equipment states, reflecting the multifaceted nature of simulation technologies. This likely arises from the fact that we can never completely recreate a version of the real world; we are always choosing parts of it to simulate.

Liu, Nikolas and Dennis (2009) have also reported inconsistencies in the definitions of fidelity, citing subjectivity and the lack of quantifying methods as key obstacles in the way of agreement. One of the earliest definitions of simulator fidelity stems from Hays (1980), who surveyed the literature and broadly classified it into two groups according to (1) *general* and (2) *specific* definitions. General definitions considered equipment reproduction by way of its accuracy and degree to which it resembled the situation (Miller, 1974), and proposed that fidelity could be achieved if simulator activity was like engaging with the actual equipment. These definitions have generally pinpointed the operational characteristics of the simulator and the degree of correspondence with the available cues, responses and actions (Miller, McAleese and Erickson, 1977). In contrast to these viewpoints, more specific definitions have emphasised the role of the environment and the task, job and behavioural performance measures in addition to physical and equipment-based similarities.

Further definitions of fidelity have considered the degree to which the environment is simulated in terms of the equipment (Zhang, He, Dai and Huang, 2007), its functionality (Allen, Hays and Buffardi, 1986), the visual and audio components (Rinalducci, 1996), the task itself (Hughes and Rolek, 2003; Roza, Voogd and van Gool, 2000), its motion cues and psychological-cognitive factors (Kaiser and Schroeder, 2003), and the simulation as a whole (Alessi, 1988; Gross, Pace, Harmoon and Tucker, 1999). Over the years, the notion of simulator fidelity has adjusted its verbiage and terminology, and perhaps mined down into a few more layers of the system, but conceptually, very little has changed. The simulation research base has grown to consider fidelity according to its physical and functional dimensions and the degree to which the individual characteristics are matched to reality. To some extent, the evolution of this literature and the tensions between general and specific definitions reflects the simulation zeitgeist that we can never completely recreate reality but we can add up all the parts to make a pretty good impression of it.

In all cases, the corpus of simulator research exhibits a trend that veers towards specificity, and as shown in Figure 8.2, considers numerous factors for fidelity that encompass the physical/environmental and psychological/behavioural components, essentially inducting the theme of skill acquisition, and evaluative measures of sensation, perception, learning and retention. Clearly, fidelity is loaded with factors that overlap and are intricately related, but the key message is that simulation incorporates fidelity in varying degrees, undertaken in an effort to convey realism.

How Much Realism is Good Enough

The question '*how much realism is good enough?*' has single-handedly propagated simulator research, and driven initiatives seeking to assess the validity and credibility of simulation-based results. Given that simulations are designed to abstract and represent different parts of reality, measurements of fidelity examine the way that they are represented. The answer to the question of how much

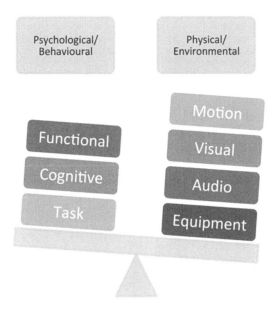

Figure 8.2   Psychological and physical constituents of simulator fidelity

realism is good enough may therefore be found in the requirements and governing criteria that guide its development. As discussed earlier, simulator fidelity may be categorised into various dimensions, including the physical, functional and environmental. Wells (1992) described *immersion, interaction* and *intuition* as three integral components essential to successfully portraying a simulator and its visual environment. These components respectively refer to the notion of the virtual environment feeling real enough to provide the experience of being surrounded by the simulation, that having an awareness that any action will manipulate the simulation, and that the simulation may also be communicated with in an obvious and altogether familiar manner.

Similarly, McMahan (2003) referred to *immersion, engagement* and *presence*, where engagement is described as a keener form of interaction that results in a deeper kind of involvement. The term presence refers to the phenomenon of behaving and feeling as if the user is in the real environment, even though they are in a simulation. Figure 8.3 consolidates the key factors and goals that contribute to the experience and sensation of simulation. All of these terms interface between the human and their impression of the simulation, but in spite of the subtle differences in what these terms mean, the effectiveness of a simulation ultimately depends on the degree to which users *perceive* it to duplicate the operational environment (Parsons, 1980). Thus, whilst realism is important for an effective simulation and generally increases with financial outlay, spending in pursuit of higher degrees of fidelity and imitation may not necessarily equate with purpose or validity.

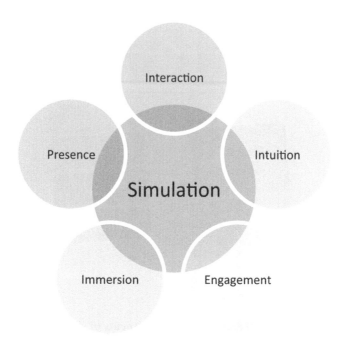

Figure 8.3   Key factors contributing to the experience of simulation

> **Making the simulated train look, feel and respond more like the actual train does not necessarily mean that people will treat it more like the actual train.**

Individual perceptions of the same simulated environment can also vary greatly, and users come to the simulator experience with a wide variety of past experiences and knowledge. Some people will be more easily and completely immersed in a simulation than others (Naweed, 2013; Naweed and Balakrishnan, 2012a). It has even been suggested that simulator fidelity can be too high. Very high fidelity may increase complexity, elevate workload and divert attention from the skills that need to be learned (Wickens and Hollands, 2000). A simulator that is perceived to exactly replicate the real train environment may result in users with unreasonably high expectations, which can negatively influence buy-in when the simulator environment (which is necessarily limited) does not meet these expectations. This is an important point, and managing expectations and attitudes towards simulator function and use in order to maximise user presence and immersion will be discussed further later in this chapter.

Not surprisingly then, the subject of high fidelity has polarised expert opinion. On one hand, early literature has praised high fidelity for maximising skills transfer (Bryan and Regan, 1972; Roscoe, 1980), and for providing performance data in meticulous detail to effectively capture errors and visual search behaviours (Fukazawa et al., 2003; Itoh, Arimoto and Akachi, 2001). Stammers and Trusselle (1999) also found that groups learning in high-fidelity simulators reported higher self-assessed confidence ratings and performed better in comparison to groups learning in low-fidelity simulators, introducing an enhanced interpersonal and subjective benefit. Thus, a high-fidelity simulator may palpably fulfil an additional role to enhance user perceptions of the organisation's visible commitment to training and safety. On the other hand, research has also indicated that *big budget* type of approaches to simulator design and implementation do not always address fidelity requirements (Hays and Singer, 1989; Lane and Alluisi, 1992; Roza, 1998). Given that different levels of skills and knowledge require different levels of fidelity, any one simulation may be unsuitable for a percentage of the population (Keon, 1990; Kinkade and Wheaton, 1972). The needs of the end users must therefore be considered systematically and early in the development programme, particularly as plausibility can be defined entirely by their requirements. Given that simulator fidelity is functionally associated with the perceptions of the human viewer, the physical accuracy in a simulation's dynamic function is arguably somewhat less important.

The user's needs and objectives define the context and purpose of the simulator, and therefore must shape fidelity requirements.

Fidelity and Skills Transfer

We have described fidelity as a core component for simulator specification but in view of the growing complexity of simulator architecture, the prolific use of simulators in various domains and an increased dependency on simulator outputs, there are obvious concerns in the way simulator fidelity is treated (Roza et al., 2000). Given that fidelity represents a key factor in the cost-benefit analysis during the early stages of simulator procurement, the concept of simulator realism requires closer scrutiny. Substantive literature has grown to dimensionalise fidelity as a sort of currency, bartering the physical, functional, environmental and psychological perspectives of simulation design into various denominations. However, the overarching aim behind all of it is to accurately identify the level of fidelity required at the outset so that the simulator may be deemed fit for purpose. The key difficulty is therefore to ensure that the technology is sufficiently accurate and convincing for the user, and as a point often missed, ensure that there is enough redundancy in the simulation to support future developments whilst limiting costs required for building, operating and maintaining.

There is much debate around the utility of simulating motion. On the one hand, it can be a key referent for perceiving acceleration, but too often the benefits are disproportionate to the costs (Young, 2003). In a simulated environment, physical motion can actually increase the potential for adverse physiological effects on the user and induce sickness (Kennedy, Lane, Berbaum and Lilienthal, 1993; Kolasinski, 1995). For human performance evaluation, one can argue that it is not incumbent to incorporate full physical cues of motion, such as yaw or roll, particularly when other features (sound vibration, semi-active seat and so on) may equally immerse the user and enhance the perception of self-motion (Rail Safety and Standards Board, 2007a). A moving base simulator is also expensive – simply adding this feature would drive the cost upwards several-fold, significantly complicate the architecture of the system and increase the chances of technical failure. Needless to say, a moving base component would also require stringent health and safety measures and regular maintenance throughout the life of the simulator. Ultimately, a motion base feature is a point of preference to those satisfied by the argument endorsing its utility but it does not suit all users, and is somewhat redundant when examining out-of-course events, equipment failures and other situations which, whilst difficult to reproduce in the real world, don't actually require motion cues (Rail Safety and Standards Board, 2007a). There is also some speculation around how well train-driving motion may be simulated.

Unlike aviation, where a moving base recreates motion according to G-forces and flight aerodynamics, replicating the algorithms underlying rail adhesion and railway physics poses a greater challenge, particularly if it is to be done realistically. Simulating train-driving motion involves simulating the vortexes created when passing trains and entering tunnels, as well as the aerodynamics associated with the train's overall form. Then there is the need to consider environmental factors and the impact of light rain or dew in order to duplicate movement associated with slippage and slide. Physical motion cues also need to consider the railway's centreline profile, the gradients and the curvature, and factor in the impact of track resistance characteristics. Last but not least, a realistic moving base should incorporate motion associated with improper train handling.

In short, there are many input vectors to consider when simulating train-driving motion. That is not to say that it cannot be done or that it doesn't enhance immersion – simulating motion is relatively easy if all that is required are the low frequency motion cues that indicate you are moving forward or going around. It is the high frequency motion data associated with the engine, the rail-wheel contact points and the couplings that have many variables and are harder to simulate. The question is, what would the user want of their motion? If it is merely for elevating immersion, then so be it, but you may encounter problems with skills transfer; if it is for developing train-handling skills, an uninformed or underrepresented model may also have implications for skills transfer. In either case, the question is, do you really need a movement base?

The motivation to use moving base simulators does not apply so readily if it is used for evaluating academic research-based applications. Academia has traditionally drawn on simulator-based methodologies that try to duplicate the underlying principles of the task and engage the user on a high cognitive and psychological plane, from simple laboratory paradigms through to full-cab replicas. In the absence of movement, motion characteristics reflective of the virtual environment have been offset by incorporating richer texture density gradients in the computer-generated imagery, changes in the ambient optic array and high-fidelity sound and vibration, providing an alternative perception of self-motion (Bruce and Green, 1990; Padmos and Milders, 1992). Thus, in both industry evaluation and academic research, fidelity may be optimised through the interface and the virtual environment.

The key to accessing high functional fidelity is to facilitate skills transfer by matching stimulus and response with the real environment.

As conceptualised in Figure 8.2, the various requirements for fidelity may be achieved by balancing them on an axis of optimum user requirements. The fidelity of the physical and equipment may vary but skills transfer will manifest when the stimulus and response (inputs and output elements) of the simulator are matched with the real environment. However, it is important to note that skills transfer can be negative as well as positive. Positive skills transfer can be defined as acquired skills which carry over to the real environment in a way that measurably improves performance and yields a good response; on the other hand, negative skills transfer relates to skill acquisition which carries over but serves only to inhibit performance in the real environment (Wickens and Hollands, 2000). Most tasks require a variety of skills that can transfer well, and if they have the same stimulus and response outputs, this can be highly positive. Conversely, negative transfer may manifest if the response elements vary from an accurate range of (simulated) stimuli.

The relationship between stimulus, response and skills transfer can be somewhat confusing and difficult to conceptualise. To help with this, Figure 8.4 analogises the relationship between stimulus and response from the rail simulation and train-driving perspective (adapted from Wickens and Hollands, 2000). Here, the stimulus (i.e. whatever the driver is reacting to) has been represented in an abstract configuration (i.e. a series of circles). The response has been represented as throttle movement. A train simulator that displays or characterises a stimulus in the same way it is found in reality (i.e. precisely the same configuration) has the potential to transfer highly positively (+ +) if the response is also the same; for example, by duplicating the exact throttle control, dashboard and response parameters of a cab into the simulator. Changing the stimulus in such a way that

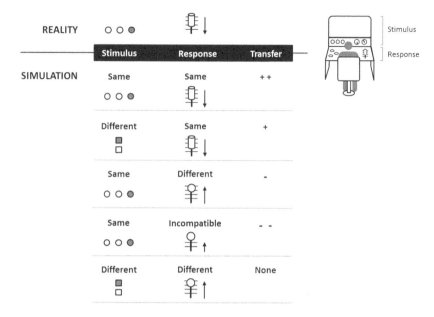

Figure 8.4   The relationship between the similarity of stimulus and response elements with transfer of training (adapted from Wickens and Hollands, 2000)

it elicits the same response also has positive transfer potential (+). For example, an individual is likely to perform well if a simulator shares the same control and response parameters, even if the dashboard happens to be different. The distinction here is that the response in the simulator needs to manifest in the same way as they do in reality, regardless of similarity or difference in the stimulus.

However, say that a simulator requires the user to push the throttle forward to decrease speed but in reality, this action is carried out using a different control and/or in a different direction. In this case, training transfer may be negative (-). This has been illustrated in Figure 8.4 as a different looking control (i.e. a knob instead of a stick) with a different direction of control (i.e. forwards instead of backwards). Skills transfer may be even more negative if the response elements bear no resemblance and are completely incompatible with one another; for example, where not only the controls for decreasing speed are different and inverted, but the actual notches in the control are also varied such that the train does not decrease speed in the same way. In this example skills transfer will be highly negative (- -). Clearly, stimulus and response issues need to be reviewed carefully when considering fidelity and skills transfer, to ensure that they map from the simulator to reality favourably. Figure 8.4 highlights that potentially negative skill transfer may be decreased if the input elements are different. For example, a driver confronted with a throttle that looks different may tolerate the transfer better if they share an identity (e.g. move in the same direction, possess the same number of grades and so on).

To revisit the moving base argument, Figure 8.4 also clarifies the potential for negative skills transfer if a motion-capable simulator is used for learning train handling but has an under-informed motion model. Offsetting a simulator's fidelity requirements is therefore ostensibly concerned with the simulator's *scope* and its intent as a tool for industry evaluation, research or both. For industry evaluation, the most salient feature for high fidelity is via the equipment and environment, with the provision of a suitable user interface (Rinalducci, 1996). The simulator would connect these spheres through the underlying structure, including (but not limited to) the track vision, audio, terrain information database, interface mock-up and system modelling. The next section will consider train driving in a little more detail, and consider the idiosyncrasies in this task that should not be forgotten when representing it in a simulation. As you read it, ask yourself, have they all been featured in the train simulators that you have used?

Simulating Train-Driving Operations

The Rail System

Rail travel is entirely land-based and is therefore constrained by the fact that it is an open system (i.e. constantly interacts with various environmental factors, such as level crossings, high traffic densities and pedestrians). However, trains are uniquely placed in that control is solely exerted through changes in speed, with the direction of movement guided entirely by track, and drivers are incapable of taking evasive action to avoid a collision. Mainstream rail operations also differ from urban light rail systems (e.g. trams) where the faster speeds and the longer, more variable stopping distances require train drivers to react to information situated outside the immediate visual field (Calvert, 2005; Rail Traffic Control, 2012) and anticipate changes in this information. Railway mechanics and infrastructure have undergone very few changes since the Victorian era, and today, in the UK and Australia, stations, bridges, viaducts and other rail constituents are being increasingly listed as heritage structures. Thus, railways are in a constant state of maintenance and upkeep, and train drivers must cope with non-routine information sources, such as speed restrictions and engineering works, which extend outside their visual field.

The level of dynamism to be found in railways instantiates the system's perceived level of *realism* and defines the sort of features that must either be simulated or controlled as part of the simulator user's experience. Figure 8.5 conceptualises a typical rail system under the tenets of a multi-system analogy (Hollnagel and Woods, 2005) and hints at the dynamic couplings in its various layers. It also illustrates the range of competing interests and levels of awareness that a train driver must learn to cope with, and which by extension, should be simulated in a training-oriented simulator. The layers of this system will be discussed in turn in the following sections.

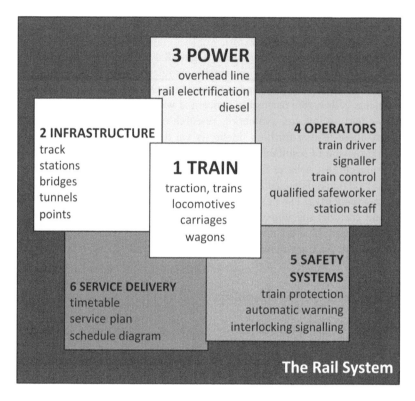

Figure 8.5   Conceptual representation of a typical rail system showing
how the various layers are dynamically coupled. Activities in
the train (1) are increasing dependent on the successive layers
(2 to 6), which create implications for control and performance
(adapted from Naweed, 2011; in press)

The Train

A train is restricted to the confines of the railway, but the interplay of the
infrastructure, such as signal changing, level crossings and point switching,
creates implications for control and performance. An accurate simulation of
a train must therefore include the means to advance and brake realistically in
view of the train-set assembly and the drag and resistance characteristics of
the track. Much like other modes of transportation, train driving is shaped by
the weather, which is uncontrollable and influences all operations (Gallardo-
Hernandez and Lewis, 2008). Thus, if weather effects are to be simulated, a
well-designed simulator would also simulate changes in track adhesion or at
least allow for its manipulation (particularly when you consider the nature
of rail traction!). But what would these values be, and how would they affect
the task?

Manipulating track adhesion in a train simulator by guessing values would introduce a fundamental confound, both in terms of the control and the realism. For example, in the case of a freight locomotive simulator, it may be that poor rail adhesion would initiate an instruction for track slippage, triggering more instances of the wheel slip indicator, which in turn would require the user to deploy sand. However, a simulator manager evaluating a new technology that displays adhesion levels may not want to control this manually and may prefer for it to be specified with greater consistency, which would require track adhesion changes to be controlled by an algorithm. Similarly, a more sophisticated simulation would incorporate the equations necessary to simulate realistic weather effects, but as discussed earlier, developing motion algorithms to match with the actual stimulus is a challenging endeavour, and indeed, there is much speculation in the rail simulation community as to whether this can be achieved at all.

> **A good understanding of the layers that comprise the rail system is important for understanding the parameters that need to inform the underlying model of a train simulator.**

To revisit the motion discussion from 'Fidelity and Skills Transfer', the argument for simulating motion in urban passenger operations is more convincing than simulating it in heavy-haul operations. Urban trains are more consistent in the ways that they handle, and driving in the urban environment is largely a combination of good train handling and route knowledge application. However, this does not apply to freight and heavy-haul train driving, which rely heavily on motion cues and feature more covert movement-based driving practices. This is because the length and loading of heavy-haul and freight trains can vary dramatically, even within a single corridor. Not only does overall size and weight come into play but also the relative distribution of the load throughout the train (e.g. heavy wagons interspersed with empty ones), as well as the number and distribution of locomotives. The interaction between these factors and track characteristics has a critical impact on train handling. Heavy-haul and freight train drivers regulate the train states of stretch (draft) and compression (buff), which are largely associated with such weight dynamics, and simulating this type of complexity, especially the motion associated with improper driving, would be extremely difficult to match with the real world.

Rail Infrastructure, Power and Operators
Beyond simulating complexity in the train layer, the dynamics of the infrastructure, power and operator layers introduce issues of traffic flow, expediency and codes of conduct that are dependent on other functional groups, such as signalling and train control. In addition to advancing and braking realistically, a basic train simulation would therefore need to incorporate the task-based parameters associated with

encountering movement authority. In reality, railways are designed to counter and control environmental impediments; the terrain, for example, may be adapted to minimise curvature, and countered through tunnels and cuttings, but unlike other ground surfaces used by transportation systems, the slightest differences in elevation of railways and their radial values yields tremendous differences in performance. This reiterates the point that traction physics must be accurately represented and introduces the issue of simulating railhead conditions.

Safety Systems

It is rare to encounter modern rail operations that do not feature safety systems based on combinations of on-board and trackside technology. These include perception-action constituents of the driving task (e.g. Automatic Warning System, Driver Vigilance Device) or a part of the overarching safety net designed to intervene in the incidence of human error (e.g. Train Protection Warning System, interlocking signalling, signal trip mechanisms). A third classification monitors speed, and if necessary, instructs error correction via the driver–cab interface (Automatic Train Protection) or an enhanced display (European Rail Train Management System). The key message here is that the parameters of any safety system, whether it is perception-action or a part of the wider safety net, should be matched against the parameters that exist in the real cab, in order to minimise any potential for negative skills transfer. Thus any inbuilt latencies found when responding to the buttons, and any intervention thresholds in the overarching safety net need to be replicated precisely. This basic premise extends to the information and visual ergonomics in enhanced displays.

Service Delivery

Balancing the goals and dilemmas between safety and performance in train driving resides entirely with the drivers – this encapsulates the core of train-driving skill and it is a skill-set that a simulator, regardless of the level of fidelity, would aim to inform and transfer to the real domain. A well-developed simulator would need to incorporate an awareness of the service delivery goals that need to be managed and sacrificed if necessary. As shown in Figure 8.6, safety and performance goals vary by operation but align with the service plan and timetabling. Heavy-haul operations have a timetabling concern as they deal with traffic management, goods transfer (minerals/grains and intermodal freight) and the driving schedule, but their key performance concern is arguably for *energy* and its conservation. On the other hand, heavy-haul safety goals are linked to the weight load, the power profile and the forces along the train, and as shown in Figure 8.6, these define a specific application of driving *form*. Of course, passenger services, both urban and suburban, also have a form concern, but given the smaller weight load, consistencies in power and manageable train forces, the key performance goal is structured around *time*. Lastly, safety in passenger operations can be defined by passenger *comfort*, both in terms of ride quality and adherence to the schedule.

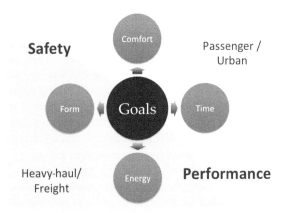

Figure 8.6   Primary train-driving goals associated with operation type and relative placement within safety and performance

> **Balancing the goals and dilemmas between safety and performance in train driving resides entirely with the drivers – this is the key train-driving skill set, and one that a simulator, regardless of the level of fidelity, would aim to inform and transfer to the real domain.**

Safety and performance are generally in conflict with one another (Naweed, Hockey and Clarke, 2013). The safety–performance issue will be extended in the task environment chapter, but suffice it to say, the complex decision-making required to walk the line between safety and performance creates tension, often pitting the two in direct competition. For example, a passenger train driver may sacrifice time if passenger comfort is compromised due to poor weather, and a heavy-haul driver may forgo energy concerns if they have a weight load that cannot be managed safely by momentum-based driving. So for train simulation, service delivery and the underpinnings of safety and performance are overarching layers of the rail system, and for simulator design, one that needs to be interrogated well in order to breed resilience in the underlying simulation model. For the driver, service delivery is informed by the basic information required in order to do their job. Thus, a train simulation needs to consider and incorporate a range of information needs necessary for real train driving, or pick and choose between them, depending on aims and attainable level of fidelity.

Train-Driving Requirements and Information Needs

In its simplest form, train driving is a dynamic task of error correction that requires the driver to correct the discrepancy between target speeds and train state. Traditionally, this is achieved by processing information from an in-depth knowledge of the infrastructure and information sources held in the environment, from signals, speed posts and other railway 'furniture', through to more natural landmarks. Clearly, misreading signals jeopardises the amount of time and distance left to stop safely and effectively, but this is somewhat confounded by the fact that information cues may be out of sight or obscured by changing conditions, necessitating an accurate awareness of the train's position at any given moment. Thus, drivers are required to perceive and process complex vestibular, kinaesthetic, acoustic and peripheral visual information to perform the task (Branton, 1979), and aligning closely with this are some very specific information needs.

Table 8.1   Summary table of basic train-driving requirements and information needs (adapted from Naweed, Bye and Hockey, 2007)

Route Knowledge	Train State Indicators	Environmental State
Gradients	Speed	Visibility
Curvature	Weight	Auditory feedback
Signals	Passenger/freight	Drag/aerodynamics
Speed restrictions	Freight type	Railhead conditions
Landmarks	Fuel usage	Weather

Table 8.1 shows some very basic information needs and driving requirements that train drivers in passenger and heavy-haul operations need in order to perform the driving task. These can generally be grouped into train state indicators, environmental state and route knowledge that is stored and processed internally but verified by available information (Naweed et al., 2007). Some of these needs are entirely informed by the rail environment and applied to a situational driving *template*. The way that an information requirement is embedded into the task is conceptualised in Figure 8.7. External information cues generally comprise route information (gradients, curvature, signals, speed restrictions, markers), and environmental conditions are largely informed by a combination of perceptual and haptic responses (based on tactile feedback). Thus, the visibility, auditory feedback, drag and aerodynamic forces around the train allow the driver to summarily profile the driving conditions. Knowing the train state allows the driver to adjust into a suitable driving style and determine the effect of environmental conditions. Knowledge of the train type, freight type and by extension, its weight, would set driving goals and inform the way

that the train is fundamentally controlled. The most dynamic train state indicator, the speed, is informed objectively by standard in-cab instrumentation and subjectively by external event rate, though this does not extend to fuel usage; the need for fuel state is more ambiguous and varies by operation type, but in the absence of in-cab indication, it is a cost-benefit parameter of driving style.

The need for route knowledge is paramount, but unlike train state indicators and environmental conditions, somewhat trickier to define. Route knowledge is driven by the dynamic interplay between a static store of knowledge and actual environmental state, enabling the driver to determine the location of the train relative to their stopping pattern and their control requirements. A stored knowledge of the route can therefore be matched against environmental content and upcoming infrastructure to prescribe movement authority (signals, speed restrictions), ascertain position (geographical markers) and update driving form (gradients, curvature). The type and range of in-cab technologies and train state indication available to the driver influences what needs to be perceived from the outside environment and how much of it, in order to apply dynamic train control. Ultimately, route knowledge is a constraint of train-driving physics and as driving skill develops, it cultivates advanced navigational strategies. How then do the layers of the rail system and the basic train-driving information needs sit within the reality of the world being simulated?

> **A train simulator needs to encourage the complex thinking related to performing safety/efficiency trade-offs.**

Figure 8.7 goes some way towards framing the physical and psychological needs for simulator fidelity. The layers comprising the rail system (see Figure 8.5), the task goals (see Figure 8.6) and the basic information needs (see Table 8.1 and Figure 8.7) paint a picture that depicts a hidden dynamic complexity, both in the railway environment and the mechanics underlying it. Train driving has clear implications for the design of a corresponding virtual environment, and a train simulator, regardless of fidelity, should accommodate the relative sophistication of the decision-making required of the task. A simulator should therefore encourage the quasi-mathematics that train drivers need to exact in order to (1) trade different driving goals, (2) functionally represent tensions in safety-performance and (3) undertake complex cognitive function to pursue dynamic train control (e.g. monitoring the environment, comprehending meaning and current state, anticipating future requirement). Clearly, this may be achieved in simulators with varying degrees of fidelity and to a certain extent the basic information needs prescribe a simple framework for simulator specification, though the choice to simulate these components are dependent on operational scope. The next section will explore how the field of train simulation has been implemented in the rail domain, starting with an overall look at the diversity inherent in simulator classification.

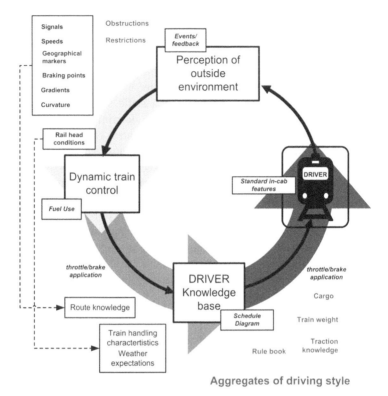

External information cues

Aggregates of driving style

Figure 8.7   Train-driving model of information needs based on traditional task requirement and standard train state features (adapted from Naweed, 2011; in press)

Train Simulators

Classes of Simulation

Generally speaking there is no accepted standard for classifying simulation, and the naming convention or nomenclature of different simulation classes varies by domain (e.g. transport), by function (e.g. driving, signalling) and by intended usage (e.g. research, evaluation). Huge diversity in the scope and depth of simulation classifications is evident (Maran and Glavin, 2003; Rehmann, Mitman and Reynolds, 1995). To some extent, the abundance of simulator nomenclature reflects the very nature of simulation – in the pursuit of faithful mimicry, simulators are powered by real-world stimuli and any attempts to classify them will merely reflect the rich variety in the real world. Classifying simulation types is therefore an arduous undertaking, though there are examples where this has been done. Clymer (1980) classified simulators into six overarching groups, including replicas, generic simulations, eclectic simulations, full-task simulations, part-task simulations and simulations that follow basic principles. As shown in Table 8.2, these classifications illustrate a range of simulation types, ranging from the complete duplicate through to the generic that focuses on analogy.

Table 8.2   Classes of simulation (adapted from Clymer (1980)

Simulator Type	Description
Replica	A complete and exact duplicate of the driver–cab interface and a very realistic projection of the environment.
Generic	Representative of a class of systems but is not a replica of any one.
Eclectic	Includes a mixture of features (e.g. manufacturer's instruments, equipment) designed to broaden the experience of the end user/ trainee.
Full-task	Deals with the entire task of the operator's command/control of the system being simulated.
Part-task	Concerned only with a part of the operator's task or operational system.
Basic principles	A generic and/or part-task simulator that omits many details in the interest of economy and simplicity, but is still capable of demonstrating the cognitive work and behaviours of the system.

Clymer's (1980) simulator decomposition recognises that a replica aims to duplicate and render all features of the real environment in order to deliver a full-simulated experience. Generic simulators aim to render the same reality, but through an environment that does not replicate the man–machine interface and thus does away with increased cost and logistical issues. Eclectic simulators recreate the man–machine interface of a real system or use a generic layout, but in an effort to stimulate user experience or garner cost-effectiveness, the integration of additional technology and features (not found in the real world) will impact overall realism. Task-based simulators distinguish the task into full- or part-based components, and depending on requirements of the task itself, may or may not include all controls or all of the features of the real environment. Basic principles simulators sacrifice the physical fidelity that will generally be found in all other categories and replace it with high psychological fidelity, forcing the behavioural responses that may be expected of the system through analogy. Stanton (1996) raised the point that such classifications should not be treated as a straight-jacketed taxonomy, as a simulator may transcend one or more descriptors, but as it stands, Clymer's (1980) decomposition does an excellent job of capturing the variety to be found in simulators.

Inconsistencies in Train Simulator Classification

As is generally the case in the area of simulation, the naming conventions in the design and use of train simulators have yet to achieve industry-wide standardisation.The range of substantive rail stakeholders (manufacturers, industry bodies, operators, researchers, etc.) has produced a train simulation nomenclature that knows no bounds and has resulted in a piece-meal approach for using descriptors. For example, a number of organisations have developed a range of train simulators for rail operators and research organisations, but each has used its own terminology to describe its products. For example, one developer has produced *Compact*, *Desk* and *Replica* train simulators, a second has developed *Portable*, *Stand Alone* and *Cab* simulators and a third has made *Full Cab*, *Driver Desk Level* and *Desktop* simulators. What do these terms actually *mean*? Is the assumption that Desk, Stand Alone and Driver Desk Level simulators belong to the same *class*? How is the term Compact distinct from Portable? Is a Cab simulator the same as a Full Cab, and furthermore do they follow the conventions of a Replica? What *does* Desk Level mean, and just *how* is this distinguished from a Desktop? At some point, one might concede that inconsistencies in train simulator classification are being exacerbated more by language than by class! Clearly, such diversity in descriptors for what is essentially the same type of product may serve only to confound procurement for rail organisations during the tendering process.

As discussed, the simulation needs of rail organisations and research institutes are ostensibly clouded by the non-standardisation in their classification systems. However, inconsistency in the use of descriptors does not reside wholly with developers but extends to government bodies. In the UK, the Rail Safety and Standards Board (2007a) classify train simulators into three distinct groups comprised of full simulators, part environment simulators and part-task trainers. On the other hand, the Federal Railroad Administration (2003) at the United States Department of Transportation classifies train simulator types into numerical grades of *I, II and III*, and furthermore, they describe all train simulators as replicas, where Type I features highest in real-world replication. Again, the question here is what *are* these denominations and how do they align with realism? Clearly, simulator descriptors need to be clarified, both to identify simulation needs and breed some much-needed consistency. Much like the definitions for fidelity given by Clymer (1980), train simulator descriptors could be clarified and classifications enhanced, first by reappraising a *general* framework, and second, by considering the domain's *specific* components.

Reconstructing Classes of Train Simulation

Establishing a General Framework

The general classes of simulation can be framed by applying a basic bottom-up process that dimensionalises in terms of the (1) *scope* of the task that's desired, and (2) the required level of *realism*. Figure 8.8 shows a general simulator classification matrix based on these referents. The scope of the task and the level of realism are base axes that underlie most simulated task environments, and engender a series of classification zones. The matrix descriptively decomposes the notion of fidelity and to some extent, consolidates Clymer's (1980) simulator classifications defined in Table 8.2. The scope of operation reflects the work or task that needs to be replicated and queries the part or full range of the task's reproduction. The level of realism reflects the depth to which the task environment needs to mimic reality in order to meet the needs of its training or research-based objectives.

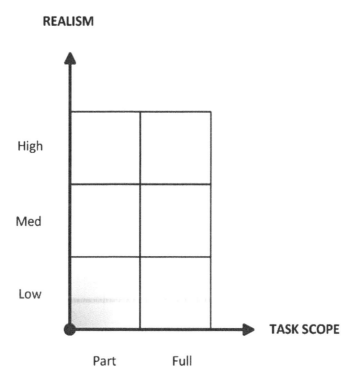

Figure 8.8   **General simulation classification matrix based on scope and realism**

In this way, replica, generic, eclectic and basic principles simulators occupy different rungs on the low to high dimension of realism. Thus, irrespective of what they are called or how they are described, all rail simulators can be constructed and mapped along this general plane. In the context of rail, a high-level taxonomy like this embraces and consolidates a variety of simulated task environments and may include functions other than train driving, such as signaller operations, trackside maintenance and fault diagnosis activities.

Defining Specific Classes
The second approach for classifying train simulation nomenclature is to use a top-down perspective to explore the existing peculiarities of the domain and consider the specific details. In the simulation framework, the task drives the context and establishing a clear set of definitions at the outset may facilitate the classification process. The *specific* train simulator classification model illustrated in Figure 8.9 consolidates the present state of rail-related simulation literature, conveys the

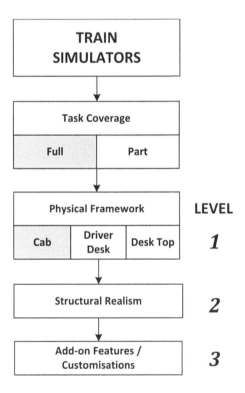

Figure 8.9   **Specific train simulator classification model showing levels of structural composition**

essence lingering behind the present treatment of descriptors and models some specific levels of simulator composition. Much like the general framework (see Figure 8.8), train simulators are filtered in terms of the full- or part-task coverage, but instead of being applied as a base axis, realism is decomposed into proponents of the physical framework, its structural realism and any additive features.

Task Coverage and Physical Framework

The simulators described in Figure 8.7 sit within two different task groupings that set the context for train simulation. Clearly, the constituents of the task may vary and be influenced by additional factors such as the type of operations (e.g. passenger, freight, heavy-haul) and training goals (procedural, route knowledge acquisition, train driving, fault finding and so on). It is worth noting that rail simulation literature is somewhat divisive in its treatment of the terms full-task and part-task. These have been used interchangeably to describe either the level of equipment fidelity in a simulator *or* the task fidelity in terms of its simulated and/or representative completeness. The trend in the treatment of these terms is a symptom of the bigger issue of non-standardisation, which has veered away from the original classification systems. In Figure 8.7, full- and part-task groupings have been used to define both the physical set of driving actions *and* the psychological state represented by its task environment. In this way, the full- and part-task definitions in Clymer's (1980) original classification system are again seen as exclusive levels in their own dimension.

The use of two task groupings contextualises the physical framework of the classification and goes some way to informing its fidelity. To formally expound and define the idea of task simulation, the term *full task* is best warranted only if the task is fully represented, and moreover, if a cab has been provisioned to isolate and enclose the environment much like it is in the real word. Given that a full-task simulator aims to represent the task in its entirety, it follows that the physical load of the task in full-task simulation is difficult to duplicate without the efficacy of an enclosed environment provided by a cab. On the other hand, a simulator that lacks an exoskeleton, for want of a better word, would necessarily be *part*-task, even if the psychological component of the task were to be fully represented. For these reasons, the specific classification model shown in Figure 8.9 uses terms that consolidate the simulator types based on the common physical framework over individual characteristics. To draw on developer terminology, we feel that the descriptors of replica, cab, full cab or the equivalent, can all be classified into a *Cab* category; desk, stand alone and desk level can all be classified into a *Driver Desk* category; and compact, desk top and portable can all be classified into a *Desk Top* category.

In Figure 8.9, the Physical Framework (Level 1) represents a dimension that gradually loses physical fidelity as it progresses from a *Cab* to the *Desk Top*. A *Cab* simulator encompasses a driver desk and the throttle/braking systems of a particular train model that are enclosed within their own environment. The internal surrounds may also incorporate additional functional equipment such as circuit breakers, security systems or supplementary communication devices, although the inclusion of these features depends on the nature of the train operation and the objectives defined for the simulator. Much like a *Cab*, a *Driver Desk* features a comprehensive driver–cab interface but is not enclosed inside a cabin. Lastly, the *Desk Top* simulator is essentially an abridged version of the *Driver Desk* that aims to represent some aspect of the train-driving task but controls and monitors are integrated into a desk with varying degrees of sophistication. Clearly, there is a large amount of variation in the structural realism (Level 2) and the add-on features or customisations (Level 3) that may be employed to convey a train simulation. However, defining the structural realism of a simulation as a function of the physical framework and any add-ons is perhaps where the area of train simulation has lost its way and yielded descriptors that have not so much expounded as confounded simulation nomenclature.

Structural Realism

The decisions that influence the implementation of structural realism in rail simulation reflect the huge diversity in the subsequent design and development of train simulators. As discussed in the previous section, the descriptors that developers use to label their products tend to describe a single characteristic of the simulator – a pattern that exhibits increased obscurity as it descends down the simulator classes. Train simulators may be better classified according to their physical framework and then according to objectives and/or contexts. Figure 8.10 extends the contents of the specific train classification model (see Figure 8.9) and draws on Clymer's (1980) simulation groupings as proponents of structural realism (see Table 8.2) into an overall schematic that models descriptors according to their aims and contexts. The driver–cab interface of the physical framework (*Cab, Driver Desk, Desk Top*) has a degree of structural realism that impacts the overall objectives of the simulator and can be grouped according to likeness with the real-world counterpart, as follows: a Replica, a Generic representation, an Eclectic mixture, and a Microworld. The next sections will refer to the structural decomposition shown in Figure 8.10 and illustrate the characteristics that engender each simulator type as an overarching function of the physical framework.

Replica
In terms of structural realism, the driver–cab interface of a Replica is indistinguishable from the driver–cab interface in the real world. Detail is matched with precision and actual equipment/component parts are used, but where this is not possible, equipment is faithfully duplicated by other means. As shown in Figures 8.11(a) and 8.11(b), a *Cab* Replica is housed in a real carbody unit or a synthetic capsule, but in either case replicates the full interior of the train cab, including the circuit breakers and switches – in other words, an experienced train driver sitting in a *Cab* Replica should not be able to distinguish the internal environment from

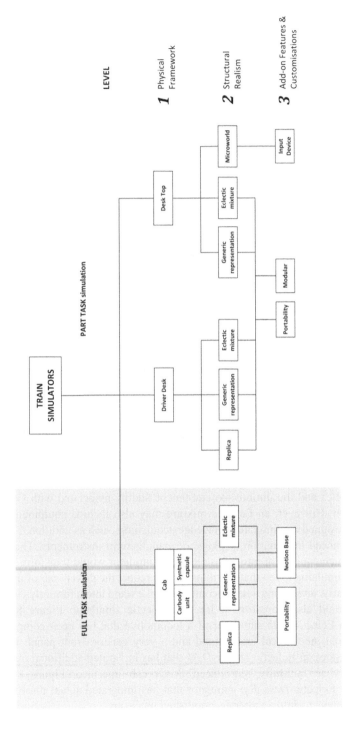

Figure 8.10 Train simulation classification diagram showing descriptors in all levels of simulator composition

a *real* cab. Large screens or image projectors are generally employed to show the driving landscape. As shown in Figure 8.12(b), a Replica for a *Driver Desk* simulator follows the same principles but invariably sheds the features relying on a full enclosure – thus, at first glance, a train driver looking at a *Driver Desk* Replica should effectively assume it has been airlifted out of a train and deposited (unceremoniously) into a room. In terms of equipment, controls in a Replica must function with the same delay and response times found in the real world. As the driver–cab interface of *Desk Top* train simulators are framed using mock-ups, they cannot be recreated in Replica form.

Generic Representation

The driver–cab interface of a Generic *Cab* representation incorporates the functional characteristics of actual equipment but not the actual equipment itself. The interior is therefore somewhat minimalistic and a composite of low-cost panelling and materials that have been structured and oriented so as to feasibly *equate* to the internals of a train cab. Input controls are not necessarily found in the same region and not all aspects of train control are simulated. Thus, whilst the simulator of a Generic *Cab* representation features a fully enclosed environment, secondary and tertiary control arms such as circuit breakers and switches are not necessarily incorporated, and primary controls need not function as precisely as a *Cab* Replica, limiting the potential for procedural learning. However, in all cases, the train-driving task is fully represented. A Generic *Driver Desk* representation adheres to the same principles minus the enclosure, whilst a Generic *Desk Top* representation [shown in Figure 8.13(a)] does so in the confines of a desk.

Eclectic Mixture

The driver–cab interface of an Eclectic simulator essentially hybridises the principles associated with the Replica and Generic representational structures. As described by Clymer (1980), an Eclectic simulator blends component parts into a composite mix, effecting a compromise between the expense and time required to create a Replica and the diminished equipment fidelity associated with Generic representation. However, an Eclectic mixture may also include equipment and/or features designed to stimulate knowledge acquisition, such as additional safety systems, enhanced information displays, or manufacturer instruments. Thus, the addition of such equipment into a base Replica simulator would transform it into an Eclectic simulator, given that it will no longer reflect the driver–cab of the real world. Similarly, integrating a real locomotive brake stand into a formerly Generic simulator would also transform it into an Eclectic simulator. Figure 8.11(b) illustrates an Eclectic *Cab* simulator of a locomotive that has been composited using an actual mechanical brake stand and a very generic front panel. Figure 8.12(b) illustrates an Eclectic *Driver Desk* that has integrated additional displays on an otherwise faithfully reproduced driver–cab interface. Figure 8.13(b) illustrates an Eclectic *Desk Top* simulator that has integrated actual throttle and braking mechanisms into a rather accomplished mock-up.

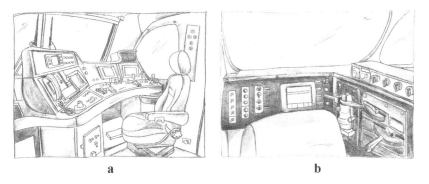

Figure 8.11 Example sketches illustrating the (a) Replica and (b) Eclectic structures in Cab train simulation

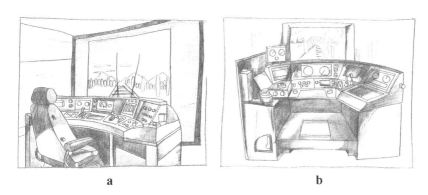

Figure 8.12 Example sketches illustrating the (a) Replica and (b) Eclectic structures in Driver Desk train simulation

Figure 8.13 Example sketches illustrating the (a) Generic Representation and (b) Eclectic structures in Desk Top train simulation

The Microworld

As shown in the classification diagram (see Figure 8.10), Microworlds are unique to the *Desk Top* class of structural realism and function by recreating the cognitive demands required of the task under simulation. Microworlds impart a task's complexity, opaqueness and dynamism via an abstract representation of its constraints and goals and a robust underlying simulation model (Brehmer and Dörner, 1993). ATREIDES (Naweed et al., 2013) is an example of one such train-driving Microworld that encourages the pursuit of safety-performance trade-offs in order to mobilise sophisticated decision-making processes, albeit in an environment with very low physical and visual fidelity. In this way, Microworlds extend Clymer's (1980) basic principles simulator classification using very different input stimuli to engage the same real-world response. As shown in Figure 8.14, Microworlds typically lack the structural realism of other *Desk Top* simulators and can be operated very easily on a stand-alone computer.

Add-On Features and Customisations

A variety of features can be added to enhance the base physical framework and structure of a train simulator without changing the classification itself. Train simulator developers overemphasise this aspect of design and even name their products as a function of these characteristics (e.g. compact, portable) though these features rarely explicate the class of simulation being advertised. As shown in the classification diagram (see Figure 8.10), a *Cab* simulator may opt for a motion or a static base and whilst this will not impact the integrity of the simulator,

Figure 8.14   Example sketch illustrating the Microworld in Desk Top train simulation

it will ultimately inform the choice of chassis. The weight and size of train carbody units render them unsuitable for a motion-base, thus in virtually all cases, motion-base simulators are designed as synthetic capsules. Similarly, carbody units may be easier to procure and offer an attractive and aesthetic proposition for rail organisations, both as an operational tool and as a showpiece, but they have very particular dimensional needs (space, placement) and would date very quickly, limiting the potential for future upgrades. An increasing number of *Cab* simulators are also being designed inside haulage containers, expanding their utility as a resource that can be driven to different training sites. *Driver Desk* and *Desk Top* train simulators can be commissioned to incorporate a range of supplementary characteristics. Modular touch screens can be used to emulate controls, gauges and (Eclectic) information displays. They may also be designed for mobility, to ease transportation to multiple training depots. Lastly, Microworld *Desk Top* set-ups may be enhanced to work with more accessible input devices or response boxes, much like that illustrated in Figure 8.14.

Examining Non-Physical Structures of Realism

This section has thus far discussed the physical framework of train simulators and considered their sub-structural elements in some detail, but clearly there are a number of constituents of structural realism that are software-based and non-physical in nature. The two main issues here are associated with the out-of-the-window display, both in terms of how it is shown and how it is generated, and the in-cab acoustics, each of which impacts the plausibility of a simulation at the higher end of fidelity.

Image Projection and the Visual Scenery

As enclosed environments, *Cab* simulators generally use projectors with screens or large flat-panel displays to show the rail environment. Either can be used for a static base *Cab* simulator, though the use of a motion base would typically require the image to be tethered to an electronic screen within the synthetic capsule. *Driver Desk* environments can also employ projectors and display screens [see Figure 8.12(a)], though they may also use flat-panel displays [see Figure 8.12(b)]. At the high end of this spectrum, developers may use multi-projection technology to create a panoramic, curved or domed 'wrap-around' view, though this would require much more space and very specific dimensions. *Desk Top* environments may utilise large screens [see Figure 8.11(b)], though at the lower end of this class would invariably favour smaller screens and desk-based display panels [see Figure 8.13(a); Figure 8.14].

The second part of this issue is concerned with software and the realism in the computer-generated imagery (CGI) of the railway environment. The data and level of detail informing CGI vary from developer to developer and improve as technology advances. It is reasonable to assume that the quality of the virtual environment would improve with financial outlay; however, given that CGI may be enhanced as a function of future upgrades, one of the most pertinent decisions for an organisation or a research institute at the outset is whether they want the virtual railway to be generic or based on their actual routes (geo-specific).

From a rail organisation's perspective, there may be greater utility in geo-specific railway simulation and an accurate representation of the track's centreline coordinates (longitude, latitude, altitude), and placement of core infrastructure (signals, railway level crossings). Failing the availability of these data, anything that allows the distances and altitude of a route to be calculated, such as the coordinates of its horizontal (x, y) and vertical (z) plane, would need to be provided. Clearly, faithful replication of track data and core landmarks from the real world would go some way to generating virtual realism and enhancing training applications; however, in the absence of substantive data, these would need to be captured and translated into data frame-by-frame, which would increase the cost of the simulator and time for deployment.

> **One of the most important decisions during simulator specification is whether the CGI of the simulator will be made up of generic or actual routes, and whether or not the capacity to construct virtual routes to the user's own specification is to be included.**

While faithful geo-specific recreation is often a primary goal, it is also important to consider the possible benefits of new route design. Having an ability to manipulate the virtual environment in order to test particular skills with a fairly high event rate may be advantageous. For example, if the goal is to assess station stopping skills, it may be more efficient to create a simulated environment with many specifically designed stations, rather than using a geo-specific simulation with station number and design dictated by the real environment. Overall, it may be optimal to purchase a simulator system that can cope with geo-specific and new routes, or to consider the possibility that more than one simulator type may be required; for example, one for evaluating a new information display and another for assessing diagnostic skills. The level of fidelity required for these different purposes may also be entirely different, and in turn, so will the initial cost, technological complexity and maintenance burden.

Sound
The simulation of in-cab noise and substantive railway acoustics goes some way to crafting the realism of a train simulator. All simulator classes will feature sound simulation, though the amount of sampling and bytes used to convey high-fidelity sound will invariably increase with budget. For *Cab* simulation, structural realism is greatly enhanced by a strategic placement of speaker equipment, and more often than not, woofers may be used to compensate for the lack of motion

(Young, 2003). In terms of the noise generated, this would include all engine noises and sound resulting from track physics, including driving over points, the '*clackety-clack*' of driving over fish plate rail joints and curve squeal. Simulating ambient background noise at stations and during specific events (passing other trains, driving over bridges and tunnels, vehicles at level crossings and so on) would also enhance the perceived realism.

Summary

In summary, train simulators can be classified according to their general framework and specific classes. The physical framework defines the overarching class, and the subcategory of structural realism characterises its context and aims. For train simulation, the descriptors offered by stakeholders appear to offer three clear classes of framework, the *Cab*, *Driver Desk* and *Desk Top*. Simulators can be commissioned within these frameworks to follow the replica, generic or eclectic structures of realism, and can also be customised for the inclusion of add-on features. It is important to note that whilst the content in this section is largely descriptive, the general classification matrix (Figure 8.8), specific classification model (Figure 8.9) and train simulation classification diagram (Figure 8.10) can also be used as a prescriptive framework to inform simulator procurement. There are many factors that must be considered during simulator procurement and most – if not *all* – are directly attributable to industry evaluation and/or research aims. The next section will consider industry and research factors, and how the end user's constraints define the most appropriate simulator for their needs.

Moving from the Physical Environment to the Task Environment

Considering End-User Constraints

Although the needs of industry and research overlap, they are frequently disparate. Yes, industry conducts research but this is rarely pursued with the same rigour for data capture and experimental manipulation that would befit an academic human factors practitioner. Thus, specification is an extremely important part of the simulator evaluation process, and unless simulator needs are carefully clarified at the outset, a developer may use default values, which will pave the way for under-specification (Naweed, 2013; Naweed and Balakrishnan, 2012b). As shown in Figure 8.15, the INPUT needs of an integrated simulator system would need to complement the needs for review and research. Industry needs would involve reviewing human performance and evaluating technology, and whilst these aims would involve a good capability for reviewing data, for advanced research (of the kind found in academia), a simulator would need more options for data capture and output. Issues around data capture and output

Figure 8.15   The INPUT level of the simulator showing how the requirements between the industry and research end users go on to shape performance measurements

format for research and review goals are discussed in the next chapter. The next section will briefly consider the levels of tractability that would befit a simulator before examining the features that end users would have to consider based on the level of their constraints.

Tractability for Research and Industry Review

The specifications for research-based simulation are very different to those for industry-based simulation. The INPUT level of the simulator will drive the focus of the next chapter, but for now, it is worth considering some basic issues of tractability for research and review. Train simulation research literature provides a very comprehensive list of features that must be carefully considered and specified when procuring a simulator for research (Yates, Sharples, Morrisroe and Clarke, 2007; Young, 2003), including:

- track building capabilities, including customisation of the route and track furnishings, railways assets and obstacles
- scenario and event programmability and manipulation of environmental conditions
- control over data acquisition rates, output resolution and format
- overall tractability.

These considerations show that researchers are interested in working with their train simulator in a way that fundamentally transcends the physical framework and even the level of structural realism. Yes these are also important factors but in the grand scheme of things, the simulator will have little research value if it does not provide executive control of data capture and output, flexibility for scenario design, capacity for future upgrading, particularly with respect to systematic research needs, and high overall tractability.

> **Unless simulator needs are carefully clarified at the outset, a developer may use default values, which will pave the way for under-specification.**

Much like researchers, industry users would significantly benefit from the option to customise their simulator scenarios and the capacity for future expansion. Although industry user constraints would generally orbit the physical fidelity dimension, electing for a Replica or Eclectic simulator may impact on future expansion or upgrading options. It may, for example, be easier to 'tack on' any new technologies or future training aids to an Eclectic simulator than it would a Replica simulator, given that the Eclectic simulator may feature this redundancy in the initial design. As touched upon in the introduction, industry and research factors govern the task environment of the simulator for the most part and will be discussed at length in the next chapter, but importantly, they also prescribe the decision-making that goes into choosing and specifying the 'right' simulator. Given the industry or research context, the overall decision-making process for procuring a train simulator will be constrained by issues associated with cost, space and usability. Of course the myriad of factors that would influence the best simulator for your needs are likely to extend beyond these (e.g. tractability, portability), but we can conceive that the moderately or highly constrained researcher and industry professional will progress through a decision tree not unlike that illustrated in the following examples.

The Moderately Constrained Industry User

The moderately constrained industry user is likely to elect for a *Cab* simulator that corresponds very closely with the internal environment of their rolling stock. Budget and space are likely to be of less concern, ruling out a Generic representation of the driver–cab interface. This user is more likely to want the Replica or Eclectic structures, though this choice will be based around their individual initiatives for training and performance review. If portability is a requirement, this user may elect to have a *Cab* simulator built inside a trailer or cargo container, facilitating ease of relocation. However, if portability is less of an issue and time to deployment is not a concern, this user may also elect for a full motion base. Ensuring that the simulator is designed with geo-specific routes would be an important consideration, and would be clarified at the outset. We may reasonably assume that the costs associated with procuring a *high-end* simulator would include extreme versatility with environmental manipulation, scenario design and track building. This is not necessarily so, and if desired, should be clarified in the specifications. Lastly, it is important that the specification addresses any short- or long-term upgrading needs associated with the hardware. Software upgrades will likely feature as part of the warranty but any options for future expansion of the driver–cab interface (to meet with changes of live equipment) would need to be specified. In summary, the key issues for this user are based around time to deployment (how quickly they want it), the level of technical complexity (how many technical problems they want to attract), maintenance issues (burden of upkeep) and review and research objectives (i.e. what they want to do with it).

The Highly Constrained Industry User

The highly constrained industry user is unlikely to have the space or the budget for a Cab simulator. This user may therefore elect for a *Driver Desk* or a *Desk Top* simulator, and again, given the needs for training and performance review, they are likely to elect for Replica or Eclectic structures. If modularity is a requirement, perhaps because multiple driver–cab configurations are required, they may also opt for a Generic representation, though this would be undesirable because of loss in equipment fidelity. A *Driver Desk* environment would be preferable, though even higher constraints or the need for portability could attract this user to an Eclectic *Desk Top*. The lower buy-in may potentially limit the simulator's aptitude for training or performance review, but this user can expect to face the same issues of geo-specific route design and scenario versatility as the moderately constrained industry user. In summary, this user would need to maximise cost-effectiveness and importantly, consider the needs of the simulator more overtly against their goals.

The Moderately Constrained Researcher

The moderately constrained researcher is likely to elect for a Generic or Eclectic *Cab* or *Driver Desk* simulator. The reasoning behind this is simple; a Replica simulator would strain the budget or otherwise limit the research potential afforded by the Generic or Eclectic structures, but given the moderate (budgetary/room) constraints, this user may opt for the added equipment fidelity gained from having a fully enclosed environment, or from a more realistic driver–cab interface. In the case of a *Cab* simulator, this user is extremely unlikely to entertain a motion base and will forgo a carbody unit in favour of adopting a synthetic capsule, though a Driver Desk is more likely to entice if portability is a concern. In either case, data control in terms of the range of performance measures that can be captured and outputted, along with formatting considerations, would be a primary concern. A moderately constrained researcher would also want to ensure that any future enhancement and core programming needs (confined to the developer), particularly those associated with iterative and systematic research needs are specified. In summary, this user would maximise cost-effectiveness, primarily ensuring that it is tractable for their research needs, but roping in some added physical/environmental fidelity for good measure.

The Highly Constrained Researcher

Highly constrained researchers come in all shapes and sizes (a good example is an impoverished PhD candidate!) and it is very unlikely that this user would have the budget, space or time to address the cost, room or technical complexity in *Cab* and *Driver Desk* simulators, regardless of their structural realism. For that reason, a high-end *Desk Top* simulator based on the Generic or Eclectic frameworks would offer this user the most value for money by way of physical fidelity. However, an extremely constrained researcher would forgo this fidelity, and use a Microworld to explore their research questions. In either case, data control, good tractability and the ease with which future enhancements and iterative changes can be implemented would be paramount.

> **Researchers are interested in working with a train simulator in a way that transcends its physical framework and level of structural realism.**

Summary

The decision-making pathways for using a train simulator, in view of particular end-user constraints, are summarised in Table 8.3. Clearly, simulator specification is not an easy task, but if the specific goals of the simulator are established early on, the deciding factors and important considerations that cannot be compromised are easier to recognise. Note that the decision-making pathway of the moderately and highly constrained researchers are essentially the same and only differ in terms of the options for physical fidelity that fewer (budgetary and space) constraints would create. Those who have a train simulator can use this as an opportunity to reflect on their procurement process. However, whilst you may have chosen the right simulator for your needs, it could be that your subsequent expectations of how it would be used have not been met. It is therefore important to consider the factors that would either make or break your capacity to use your simulator optimally.

Table 8.3   Summary table of simulator specification criteria and decision-making pathways for moderately and highly constrained users

End-user constraints			Industry		Research	
			Moderate	High	Moderate	High
Physical framework and structural realism	*Cab*	Replica	x			
		Generic			x	
		Eclectic	x		x	
	Driver Desk	Replica		x		
		Generic			x	
		Eclectic		x	x	
	Desk Top	Generic				x
		Eclectic		x		x
		Microworld				x
Deciding factors		Motion base	x			
		Portability	x	x	x	x
		Modularity		x		
Other important considerations		Geo-specifics	x	x		
		Scenarios	x	x	x	x
		Upgrading	x	x	x	x
		Tractability			x	x
		Data control			x	x

Conclusions

Taken together, this chapter has described the various features that make up the physical environment of the integrated simulator system. These features frame the simulation but are themselves encased by fidelity, which demands that the physical/environmental and psychological/behavioural constituents of the real world being replicated are adequately matched. For train driving it is important to ensure that the idiosyncrasies and peculiarities of the rail domain, which instil very particular information requirements, are effectively conveyed and represented in the task. To this end, the degrees of fidelity, and the ways in which the driver–cab can be reconstructed, create many incarnations for train simulation that would cater for the physical needs of the end user. However, simulator technology is growing very rapidly, and despite the advancements in realism and mimicry, train crew training managers and academic researchers are facing the same question – how do we *wield* these technologies?

There are many parts to fidelity, many classes of simulator and even more distinctions of structure, necessitating more levels of simulator shrewdness than ever before. Despite the obvious polarity between their methods and goals, simulator users in both industry and research want to be able to harness these technologies meaningfully and get better cost-effectiveness. This chapter has explored the issue as it applies to both these users and analogised fidelity into a currency that relies on a good understanding of its applications and inherent limitations to prove lucrative. There is little doubt that if used appropriately, train simulators may advance skills acquisition and enhance systematic research, but the glamour and polish in the nomenclature alone must be traversed before the actual equipment can be confronted.

> **Using a simulator can be rewarding but harder to control for than the real world.**

Train simulators will continue to be developed and marketed under obfuscating labels and a variety of names, but all will operate under the basic classification system detailed in this chapter – the *Cab*, the *Driver Desk* and the *Desk Top*. A good understanding of the relative strengths and weaknesses of these systems and of the costs and resource requirements is needed to make an informed decision and not get caught up in the terminology. To that end, the distinctions in structural realism that essentially reframe work from the 1980s identify the kind of simulator that would work best.

For the industry user, it is important to realise that high physical fidelity is not the panacea that people think. There is speculation around whether present technology can replicate train-driving motion faithfully, and whilst positive testimonials for motion in urban operations are by no means uncommon, there are precious few for heavy-haul. For now, a motion base may just be a way of encouraging the user to accept the train simulator, rather than using it to overtly evaluate train-driving skills. Tractability for the user's need is really the key – remember, a simulator can be harder to control than a real train. Thus, it is important that your simulator specifications meet your needs head-on. Simulator fidelity is an objective issue bound within the subjective experience, and one that for train simulation may not be answered to everyone's satisfaction. However, there is little doubt that if wielded appropriately, the train simulator is a tool that paves the way for effective evaluation of technologies, enriches the experience of skill learning and supports a more comprehensive understanding of the influence of operating conditions on the train driver.

Acknowledgements

The authors gratefully acknowledge the assistance of Richard Bye (Network Rail, UK) and Frank Hussey for their input and helpful advice on early conceptualisations of some content in this chapter. The authors would also like to thank a number of key contacts in Australian rail organisations who provided access to simulators and/or helped inform the content of this chapter, including Louise Tsagaris, Gregory Reczek, Ron Devitt and Garth Schwartz (RailCorp), Dean Pickett and Bernie Perry (Public Transport Authority, Western Australia), Adrian Hurley, Gary Bradshaw and Paul Bashford (QR National), and Max Atkinson (Department for Planning, Transport and Infrastructure, South Australia).

References

Alessi, S.M. (1988). Fidelity in the design of instructional simulations. *Journal of Computer Based Instruction, 15*(2), 40–47.

Allen, J.A., Hays, R.T., and Buffardi, L.C. (1986). Maintenance training simulator fidelity and individual differences in transfer of training. *Human Factors, 28*(5), 497–509.

Branton, P. (1979). Investigations into the skills of train-driving. *Ergonomics, 22*(2), 155–64.

Brehmer, B., and Dörner, D. (1993). Experiments with computer-simulated microworlds: Escaping both the narrow straits of the laboratory and the deep blue sea of the field study. *Computers in Human Behavior, 9*(2–3), 171–84.

Bruce, V., and Green, P.R. (1990). *Visual perception: physiology, psychology and ecology.* Array Hove: Erlbaum.

Bryan, G.L., and Regan, J.J. (1972). Training systems design. In H.P.V. Cott and R.G. Kinkade (eds.), *Human engineering guide to equipment design* (pp. 667–99). Washington, DC: United States Department of Defense.

Calvert, J.B. (2005). *Railways: History, signalling, engineering.* Retrieved 23 December 2011 from http://mysite.du.edu/~jcalvert/railway/railhom.htm.

Clymer, A.B. (1980). Simulation for training and decision-making in large-scale control systems. *SIMULATION, 35*(2), 39–41.

Cullen, W.D. (2001). *The Ladbroke Grove Rail Inquiry Part 2 Report.* Suffolk: HSE Books.

Federal Railroad Administration. (2003). CFR 49 *PART 240 – Qualification and Certification of Locomotive Engineers.* Washington, DC: United States Department of Transportation, Federal Railroad Administration.

Fukazawa, N., Kuramata, T., Satou, K., Sawa, M., Mizukami, N., and Akatsuka, H. (2003). Human Errors Committed on a Train Operation Simulator. *RTRI Report (Railway Technical Research Institute), 17*(1), 15–18.

Fukazawa, N., Kuramata, T., Satou, K., Sawa, M., Mizukami, N. and Akatsuka, H. (2004). Acquisition process of speed control skill in operating a train-driving simulator. *RTRI Report (Railway Technical Research Institute), 18*(2), 35–40.

Gallardo-Hernandez, E.A., and Lewis, R. (2008). Twin disc assessment of wheel/rail adhesion. *Wear, 265*(9–10), 1309–16.

Gray, W.D. (2002). Simulated task environments: The role of high-fidelity simulations, scaled worlds, synthetic environments, and laboratory tasks in basic and applied cognitive research. *Cognitive Science Quarterly, 2*(2), 205–27.

Gross, D.C., Pace, D., Harmoon, S., and Tucker, W. (1999). Why fidelity? *Proceedings of the Spring 1999 Simulation Interoperability Workshop, 14–19 March 1999, Orlando, Florida.*

Hays, R.T. (1980). *Simulator Fidelity: A Concept Paper* (Technical Report 490). Alexandria, VA: DTIC.

Hays, R.T., and Singer, M.J. (1989). *Simulation Fidelity in Training System Design: Bridging the Gap Between Reality and Training.* London: Springer-Verlag.

Hollnagel, E., and Woods, D.D. (2005). *Joint Cognitive Systems: Foundations of Cognitive Systems Engineering.* Boca Raton, FL: Taylor & Francis.

Hughes, T., and Rolek, E. (2003). Fidelity and validity: issues of human behavioral representation requirements development. *Proceedings of the 2003 Winter Simulation Conference, 7–10 December 2003, New Orleans, Louisiana.*

Itoh, K., Arimoto, M., and Akachi, Y. (2001). Eye-tracking applications to design of a new train interface for the Japanese high-speed railway. In M.J. Smith, G. Salvendy, D. Harris and R.J. Koubek (eds.), *Usability Evaluation and Interface Design: Cognitive Engineering, Intelligent Agents and Virtual Reality* (vol. 1, pp. 1328–32). Mahwah, NJ: Lawrence Erlbaum Associates.

Kaiser, M.K., and Schroeder, J.A. (2003). Flights of fancy: the art and science of flight simulation. In M.A. Vidulich and P. Tsang (eds.), *Principles and Practice of Aviation Psychology* (pp. 435–71). Mahwah, NJ: Lawrence Erlbaum Associates.

Kennedy, R.S., Lane, N.E., Berbaum, K.S., and Lilienthal, M.G. (1993). Simulator sickness questionnaire: An enhanced method for quantifying simulator sickness. *International Journal of Aviation Psychology, 3*(3), 203–20.

Keon, L. (1990). Role of simuators in the railway industry: Human factors - trainees and trainers. *Proceedings of the 2nd Seminar in Transportation Ergonomics, 24–25 October 1990, Montreal, Quebec, Canada.*

Kinkade, R.G., and Wheaton, G.R. (1972). Training device design. In H.P.V. Cott and R.G. Kinkade (eds.), *Human Engineering Guide to Equipment Design* (pp. 667–99). Washington, DC: United States Department of Defense.

Kolasinski, E.M. (1995). *Simulator Sickness in Virtual Environments* (Technical Report 1027). Alexandria, VA: DTIC.

Lane, N.E., and Alluisi, E.A. (1992). *Fidelity and Validity In Distributed Interactive Simulation: Questions and Answers* (IDA Document D-1066). Alexandria, VA: DTIC.

Liu, D., Nikolas, M., and Dennis, V. (2009). Simulation fidelity. In D.A. Vincenzi, J.A. Wise, M. Mouloua and P.A. Hancock (eds.), *Human Factors in Simulation and Training* (pp. 61–73): Boca Raton, FL: CRC Press, Taylor and Francis Group.

Maran, N.J., and Glavin, R.J. (2003). Low- to high-fidelity simulation – a continuum of medical education? *Medical Education, 37*, 22–8.

McMahan, A. (2003). Immersion, engagement and presence: a method for analyzing 3-D video games. In M. Wolf and B. Perron (eds.), *The Video Game Theory Reader* (pp. 67–86). New York: Routledge.

Miller, G.G. (1974). *Some considerations in the design and utilization of simulators for technical training* (AFHRL-TR-74-65). A.F.H.R. Texas: Laboratory.

Miller, L.A., McAleese, K.J., and Erickson, J.M. (1977). *Training Device Design Guide: The Use of Training Requirements in Simulation Design* (F33615-76-C-0050). A.F.H.R. Laboratory.

Naweed, A. (2011). *Enhanced information design for high speed train displays.* (Doctoral dissertation). University of Sheffield, UK.

Naweed, A. (2013). Simulator integration in the rail industry: the Robocop problem. *Proceedings of the Institution of Mechanical Engineers, Part F: Journal of Rail and Rapid Transit.* doi: 10.1177/0954409713488365.

Naweed, A. (in press). Investigations into the skills of modern and traditional train driving. *Applied Ergonomics.*

Naweed, A., and Balakrishnan, G. (2012a). Perceptions and experiences of simulators as a training tool in transport: The case of the Australian rail industry. *Road & Transport Research: A Journal of Australian and New Zealand Research and Practice, 21*(3), 77–85.

Naweed, A., and Balakrishnan, G. (2012b). That train has already left the station! Improving the fidelity of a freight locomotive simulator at post-deployment. In E. Leigh (ed.), *Conference Proceedings of the Asia-Pacific Simulation Technology & Training Conference (SimTecT) 2012, 18–21 June 2012.* Adelaide: Simulation Australia.

Naweed, A., Bye, R., and Hockey, G.R.J. (2007). Enhanced information design for high-speed train displays. In D. de Waard, J. Godthelp, F.L. Kooi and K.A. Brookhuis (eds.), *Human Factors Issues in Complex System Performance* (pp. 235–40). Maastricht: Shaler Press.

Naweed, A., Hockey, G.R.J. and Clarke, S.D. (2013). Designing simulator tools for rail research: The case study of a train driving microworld. *Applied Ergonomics, 44*(3), 445–54.

Padmos, P., and Milders, M.V. (1992). Quality criteria for simulator images: A literature review. *Human Factors, 34*(6), 727–48.

Parsons, H.M. (1980). *Aspects of a Research Program for Improving Training and Performance of Navy Teams.* Alexandria, VA: DTIC.

Rail Safety and Standards Board. (2007a). Good Practice Guide on Simulation as a tool for training and assessment *RS/501 Issue 2.* London: Rail Safety and Standards Board.

Rail Safety and Standards Board. (2007b). Good Practice in Training. A guide to the analysis, design, delivery and management of training *RS/220.* London: Rail Safety and Standards Board.

Rail Traffic Control. (2012). *Conventional control techniques.* Retrieved 15 January 15 from http://www.britannica.com/EBchecked/topic/601854/traffic control/64284/Rail-traffic-control.

Rehmann, A.J., Mitman, R.D., and Reynolds, M.C. (1995). *A Handbook of Flight Simulation Fidelity Requirements for Human Factors Research* (DOT/ FAA/CT-TN95/46). Wright-Patterson AFB, OH: Crew System Ergonomics Information Analysis Center.

Rinalducci, E. (1996). Characteristics of visual fidelity in the virtual environment. *Presence, 5*(3), 330–45.

Roscoe, S.N. (1980). *Aviation Psychology.* Ames: Iowa State University Press.

Roza, M., Voogd, J., and van Gool, P. (2000). Fidelity considerations for civil aviation distributed simulations. *Symposium at the AIAA Modeling and Simulation Technologies Conference, 14–17 August 2000, Denver, Colorado.*

Roza, Z.C. (1998). *Simulation fidelity: A preliminary study to fidelity assessment methodologies for HLA based distribution simulations (FEL-98-1311).* The Hague: T.P. a. E. Laboratory.

Stammers, R.B., and Trusselle, T.L. (1999). The effect of reduced simulator fidelity on learning and confidence ratings. In M.A. Hanson, E.J. Lovesey and S.A. Robertson (eds.), *Contemporary Ergonomics* (pp. 453–7). London: Taylor & Francis.

Stanton, N.A. (1996). Simulators: a review of research and practice. In N.A. Stanton (ed.), *Human Factors in Nuclear Safety* (pp. 114–37). London: Taylor & Francis.

Wells, M.J. (1992). Virtual reality: technology, experience, assumptions. *Human Factors Society Bulletin* (35), 1–3.

Wickens, C.D., and Hollands, J.G. (2000). *Engineering Psychology and Human Performance (3rd edn).* Upper Saddle River, NJ: Prentice-Hall.

Yates, T.K., Sharples, S., Morrisroe, G., and Clarke, T. (2007). Determining user requirements for a human factors research train driver simulator. In J. Wilson, B. Norris, T. Clarke and A. Mills (eds.), *People and Rail Systems: Human Factors at the Heart of the Railway* (pp. 155–65). Aldershot: Ashgate Publishing.

Young, M.S. (2003). Development of a railway safety research simulator. *Contemporary Ergonomics,* 364–9.

Zhang, L., He, X., Dai, L.-C., and Huang, X.-R. (2007). The simulator experimental study on the operator reliability of Qinshan nuclear power plant. *Reliability Engineering & System Safety, 92*(2), 252–9.

Chapter 9

Evaluating Your Train Simulator Part II: The Task Environment

Jillian Dorrian

School of Psychology, Social Work and Social Policy,
University of South Australia, Adelaide, Australia

Anjum Naweed

Central Queensland University, Appleton Institute, Adelaide, Australia

Introduction

In the previous chapter, simulation design and fidelity were discussed, with the conclusions that the objectives for a simulator are important to identify during its specification. This chapter will expand on this idea, but while the focus of the previous discussion was on the design and classification of the simulator environment, this chapter will focus on the data that are recorded by simulators and how they may be used in view of the simulator's objectives. In broad terms, train simulators are typically designed for three key purposes, as follows:

- Driver training
- Competency assessment
- Research

Using simulators to deliver training allows familiarisation with routes, certain track features and responses to emergency situations in a relatively consequence-free setting. Using them to assess competency enables the ongoing measurement of performance. Finally, using them for research enables the systematic investigation of the effects of differences at the driver (e.g. mood, experience, memory, fatigue), train (e.g. type, length, loading, locos, mechanical malfunctions), track (e.g. curvature, length, grade, defects, works, crossings) and environmental (e.g. light, noise, visibility, ice, rain, snow) levels on driving performance. As shown in Figure 9.1, a wide range of factors influence the ability to drive a train, highlighting an information-rich environment. However, simulation requirements, particularly the type and detail of the data required for training, competency and research, are often at odds with one another. The previous chapter made the point that simulators can be used to evaluate more than just technology, and that the fabric of the simulator must also be evaluated if meaningful data are to be derived

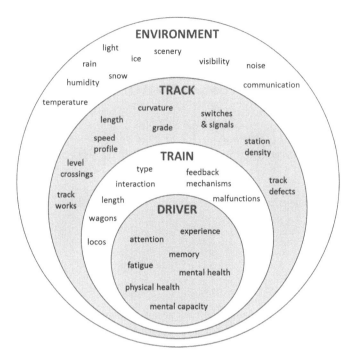

Figure 9.1   **Factors that can influence train-driving performance from inside the cab to the outside environment (adapted from Dorrian, Roach, Fletcher and Dawson, 2007)**

from the evaluations of humans, systems and the processes placed inside. This chapter will extend this notion but focus more overtly on how these evaluations may be performed for research and review, with a focus on data format, choice of measurement, and interpretation and analysis of data.

A simulator is an information presentation and performance monitoring *system* made up of many factors designed to aid in the understanding of train-driving performance. In order to develop this system, we need a specific idea of what the purpose of the simulations are, what we aim to measure and how we aim to measure it, focusing on:

- the needs of all potential groups who may use the simulator (trainees, experienced drivers undergoing performance review or learning new routes or techniques, researchers, etc.)
- the best way to capture the task of driving a train – the underlying concepts behind train driving and the different aspects of the driving task
- the range of factors that influence the ability to drive a train (Figure 9.1), including knowledge of which factors would be useful to control or manipulate for training, review or research

- different ways to measure and collect train-driving performance (e.g. speed, brake use, throttle use, fuel consumption, buff or draft forces, station stopping, on-time running)
- the way the data should be collected, stored and outputted in order to meet the requirements of all user groups.

These areas will be considered in turn throughout this chapter. A visual representation of the system is displayed in Figure 9.2, and as with any working system, consideration is required at the *input, processing* and *output* levels. The inputs of the system are the previously discussed demands of training, performance review and research. We need to be keenly aware of the end users of the simulator system and the ways in which they require the simulator to function. This will be discussed further below in 'The Simulator System INPUT: Technologies for Training, Review and Research'.

At the processing level, the performance measures collected in the simulator must be chosen very carefully to give a complete picture of the way in which performance is changing. This is particularly important given that train driving is actually a very complex task to perform. As shown in Table 9.1, a train driver engages in multiple simultaneous tasks during any one point of a journey, each of which has its own cognitive demand, and the different factors at the driver, train, track or environmental levels may influence some or all of these cognitive demands in different ways, and may affect performance measures in different ways.

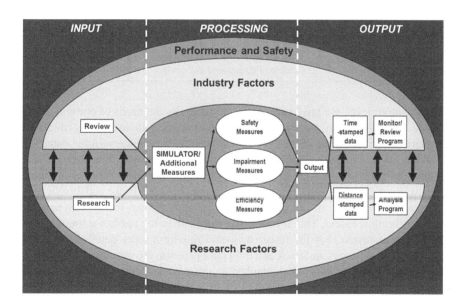

Figure 9.2   Elements of an integrated simulator system for measuring performance

Thus, if performance measures are not chosen carefully, the effects of a certain factor on train-driving performance may be missed, or the overall influence of that factor on performance may be misinterpreted. This will be illustrated using the example of a driver who is fatigued in 'The Simulator System PROCESSING: Specifying and Interpreting Train-Driving Measures'.

Table 9.1   Train-driving tasks, associated cognitive demands and some examples of performance measures (suggested definitions and units for these measures and others are given in Table 9.3)

Train-driving tasks (Grant, 1971)	Cognitive demands (Roth, 2000)	Performance measures
1. Determine track position 2. Estimate distance to next track feature 3. Combine (1 and 2) with train knowledge (e.g. locos, length, loading) and track knowledge (e.g. speed restrictions, track works)	• Recall track information • Possess and maintain situation awareness • Detect and recognise objects in or around the right-of-way • Detect and recognise violations of expectations • Monitor cab information • Monitor radio communication • Maintain sustained attention • Coordinate and cooperate with others • Formulate appropriate responses to situations • Plan and make decisions • Establish priorities and manage workload/attentional demands.	• Fuel use • Brake use • Throttle use • Trip time • Buff and draft forces • Penalty brake applications • Speed violations • Station stopping accuracy • Signal Passed at Danger (SPAD)

At the output level, we must again consider carefully the needs of our end users. While simulators may be used for driving practice and instructing different scenarios, in many cases, the users will also want to save and revisit the data. For example, driver trainers may need the output to be exported to a program that allows the data to be visualised simply and clearly for immediate performance review. Such programs may generate basic graphs of speed, brake and throttle use, and inter-train forces, in order to examine driver responses to different aspects of the driving scenario. This level of data may satisfy the needs for training and performance review, but for research, would require a more detailed record of performance measures and track and train parameters in the simulated drive. They will typically require data to be outputted to a data analysis program for statistical inquiry. Therefore, data output should be in a form that can be adapted, easily extracted, analysed and interpreted. This will be explored in more detail in 'The Simulator System OUTPUT: Data Sampling and Format'.

The Simulator System INPUT: Technologies for Training, Review and Research

In the past 20 years, there have been significant advances in rail performance-monitoring technologies such as simulators. These technologies have been primarily designed for training or competency review (e.g. quality control, incident investigation) but are being increasingly used for driver training, focusing on driving technique, familiarisation with new corridors, testing new in-cab technologies and training for critical rail events (Itoh, Arimoto and Akachi, 2002; Tichon, Wallis and Mildred, 2006). In general, use of these technologies for research purposes has never been a high priority, and there has been a lack of dissemination of experiences, difficulties and findings from the ways in which simulators have been used (which we hope this book will help to address).

> **Use of simulators for research purposes has often been an afterthought, rendering their application limited, difficult and time-consuming.**

From a research perspective, laboratory studies provide a controlled environment in which to test the specific effects of manipulating certain effects of interest (these are *independent* variables). However, as laboratory studies are unable to incorporate all of the variables that may intervene in a workplace, findings are often limited in operational relevance. Although field studies have higher intuitive appeal due to greater face validity (i.e. they look like they will measure what they are supposed to measure), they are usually limited by the constraints of the work environment (including access to participants, time, cost, resources, political issues) and uncontrolled contributory factors. To reiterate a key point from the previous chapter, simulator studies synthesise laboratory and field approaches to research, enabling a greater ability to recreate workplace tasks while maintaining a degree of control over potential confounds (e.g. consistency in scenario conditions). They also allow the planned investigation of potentially hazardous circumstances (e.g. mechanical defects, severe weather conditions, impaired driver) in a safe and ethical way.

Table 9.2 displays the kind of operational and research goals that could be addressed with an appropriately designed simulator system and some considerations regarding the simulation that may accompany these goals. For example, for training and performance review, it may be preferable to have the capacity to accurately simulate train, track and trackside features so that they match the actual track, or match aspects of the actual track as closely as possible. This would be important if the simulator is to be used for track knowledge acquisition (i.e. assisting in developing awareness and memory of track features such as stations, crossings, grade and so on). Accuracy of trackside features (e.g. houses, buildings, signs, trees) is often considered to be of lower importance relative to track and train features, however, since drivers may use such features as cues for train control (e.g. 'I apply the air brake when the front of the train reaches the house with the white picket fence'), it may be beneficial to reproduce the trackside environment as accurately as possible, or at least consider accurate simulation of a broader range of features which may be used as key driving cues.

Table 9.2   Goals for training, review and research and some considerations for the simulation

Inputs	The simulator could be used to ...	Accuracy of virtual train handling	Accuracy of train and track features	Accuracy of trackside features	Ability to manipulate train and track features	Ability to manipulate trackside features	Ability to create hazardous conditions circumstances	Detailed data output	Data output for fast review
Operational	teach or review the basics of driving	x	x	x					x
	teach or review knowledge of new routes	x	x	x					x
	teach or review driving techniques during hazardous conditions	x	x	x	x	x	x		x
	teach or review track rules	x	x	x					x
	review circumstances surrounding an incident or accident	x	x	x			x	x	x
	teach or review driver training practices and procedures	x	x	x			x	x	x
Research	investigate the effects of changes at the *driver* level (e.g. experience, mood, fatigue, workload) on performance				x	x	x	x	
	investigate the effects of changes at the *train* level (e.g. length, loading, locos) on performance				x	x	x	x	
	investigate the effects of changes at the *track* level (e.g. curvature, grade, speed limits) on performance				x	x	x	x	
	investigate the effects of changes at the *environment* level (e.g. weather, slip, visibility) on performance				x	x	x	x	

In contrast, for research purposes, it may be more advantageous to have the ability to manage and manipulate the train, track and trackside features, rather than focusing on the realism. In order to investigate any effects on performance, data output requirements are also likely to be more detailed and less focused on immediate review. Researchers may also wish to have some control over the immediate history of the research participants, and various factors to which they may be exposed, that could have an impact on performance. For example, performance researchers are acutely aware of time-of-day fluctuations in performance (often referred to as our circadian performance rhythm), which results in poorer performance at night, particularly between the hours of 3 and 6am (Rajaratnam and Arendt, 2001). Thus, when comparing the effects of various experimental manipulations on performance, researchers may strive to keep the time of day constant (e.g. investigating the effects of sleep loss on train driving by comparing two 9am test sessions in the simulator, one following a night of sleep and one following a full night awake). Researchers may also wish to control, or at least measure, intake of alcohol, caffeine, sugar, calories, sleep, stress, sunlight exposure, temperature, social contact, work history, age, driving experience, hand dominance, visual acuity – among a plethora of factors that can influence performance – in order to ensure that they are measuring the effects of the experimental parameter of interest on the ability to drive a train.

> **An accurate high-fidelity representation of the work environment may be a primary concern in an operational context, whereas for research, a larger issue tends to be the ability to control and manipulate aspects of the simulator environment.**

The Simulator System PROCESSING: Specifying and Interpreting Train-Driving Measures

Given the complexity of the train-driving task, it is important to select performance measures strategically, so that they cover a wide range of the task's cognitive demands. The first step in doing this is to make sure that our performance measures are clearly defined

Defining Measures

Table 9.3 provides some examples of performance measures, definitions and units of measurement, as used in a number of train studies involving simulators and data loggers (Dorrian, Hussey and Dawson, 2007; Dorrian, Roach, Fletcher and Dawson, 2007; Dorrian et al., 2006; Roach, Dorrian, Fletcher and Dawson, 2001). This list is by no means exhaustive, and suffice it to say, there are a plethora of measures that may be collected and many ways in which brake use, throttle use, speed, forces and violations can be considered. Each of these measures may prove more or less sensitive to the effects that the researcher or driver trainer is trying to capture. Choosing less sensitive measures may result in the (misleading) impression that a particular factor has no effect on performance, while choosing only the most sensitive measures may exaggerate the importance of the effect. Similarly, choosing only certain types of measures may lead to a misunderstanding about the way in which certain factors act on performance. This is illustrated clearly in the following example, where simulators were used to investigate the effects of fatigue on train-driving performance.

Table 9.3   Definitions and units for train-driving measures and violations

Parameter	Way of measuring it	Definition	Units
Fuel	Fuel use	Total amount of fuel used	L
	Fuel rate	Litres of fuel used per km travelled per gross tonne of train weight	L/km/GT
Brake	Brake use	Air brake applications, as indicated by brake pipe pressure (BPP)	kPa
	Heavy brake reduction	A reduction of >70 kPa or >100 kPa (depending on threshold chosen) with a train speed of >20 km/h	frequency
	BPP over-utilised	Brake pipe pressure drops below 360 kPa	frequency
	No split service	Brakes applied again <1.5 s after an initial reduction has been made	frequency
Throttle/ Dynamic Brake	Throttle use	Throttle applications, as indicated by throttle position	position (1–8)
	Throttle changes	Number of changes in throttle position	frequency
	Fast throttle changes	Throttle position changes occurring within 3 s of each other	frequency
	Dynamic brake use	Time spent in dynamic brake	min
Speed	Average speed	Mean speed across the trip	km/h
	Maximum speed violation	Time spent over maximum train speed (e.g. 110 km/h)	min
	Speed violations	Exceeding the speed limit for the track, often classified as: Mild, >10% above limit; Moderate, 10–25% above limit; Extreme, >25% above limit	frequency
	Speed error	Deviation from pre-planned optimal journey trajectory using root mean square values	km/h
	Trip time	Time taken from origin to destination	min

Forces	Buff and draft forces	Forces along the train when wagons push together (buff) and stretch apart (draft)	kN
	Passenger discomfort	Abrupt changes in acceleration (km/h) over time (s) exceeding 2 m/s^2	frequency
Other violations	Penalty brake applications	Failure to acknowledge the in-cab vigilance system in a timely fashion results in an emergency brake application, bringing the train to a full stop	frequency
	SPAD/limit of authority breach	Exceeding stop signal/limit of authority into unauthorised track	frequency
	SPAD severity	Distance of SPAD	m
	Independent brake violation	Independent brake application while either the throttle or dynamic brake are engaged	frequency

Choosing Measures to Fully Capture Performance Effects: A Fatigue Example

'Quite often you'll be driving and think to yourself, "Did I stop at that last station?" You can't bring it back. You think "Geez I hope I did."' p. 4
'Your breaking patterns change when you're tired; it's harder to pull up a loco.' p. 4
'I'll start to see shadows or blow the horn in weird places.' p. 4
'If we all gave in to fatigue the [transport organisation] couldn't run. It's just part of life to be tired.' p. 3
(Quotes about sleep loss and fatigue from train drivers in Thompson, Rainbird and Dawson [2010])

As portrayed in the above quotes from train drivers, there is mounting evidence to show that sleep loss and fatigue can compromise driving efficiency and safety (Table 9.4). Since the first study of human performance and sleep deprivation in 1896, research has demonstrated an association between sleep loss and impaired performance (Patrick and Gilbert, 1896). Significantly however, not all aspects of performance are affected by fatigue in the same way. The extent to which task performance is impaired is dependent on a task's underlying parameters, including duration (Dinges and Powell, 1988), complexity (Lamond and Dawson, 1999) and interest (Johnson, 1982). As such, findings regarding the type and magnitude of performance impairment during sleep deprivation are, to a large extent, contingent on the particular parameter(s) being tested in each study. As discussed above, this applies to the way that we choose to measure train-driving performance, and therefore, thinking through the measures and selecting them carefully is important. Given the complexity of the train-driving task, an important question for performance researchers is – how exactly is train-driving performance affected by fatigue?

Table 9.4   Some examples of causes and effects of fatigue in rail

Causes of fatigue	Effects of fatigue
• Lower sleep quality and quantity • Uncertain shift start times • Long wait times • Long commutes • Sub-optimal sleeping conditions at terminals • Lack of rest/napping prior to night shift	• Drowsiness • Microsleeps or 'sleep attacks' while driving • Increased fuel use • Increased speed violations • Failure to sound the horn at crossings • Rail accidents
(Dorrian, Baulk and Dawson, 2011; Foret and Latin, 1972; Härmä, Sallinen, Ranta, Mutanen and Muller, 2002; Pilcher and Coplen, 2000; Pollard, 1991)	(Åkerstedt, Torsvall and Froberg, 1983; Cabon, Coblentz, Mollard and Fouillot, 1993; Dorrian, Hussey et al., 2007; Dorrian, Roach et al., 2007; Kogi and Ohta, 1975; Lauber and Kayten, 1988; Pearce, 1999; Torsvall and Åkerstedt, 1987)

The Effects of Fatigue on Simulated Driving Performance

An Australian rail simulator study (Dorrian, Roach et al., 2007) investigated the performance of 20 freight train drivers at different levels of work-related fatigue during a 432-km track section in Queensland (trip time was approximately 100 min). Driving performance was measured at different points in the drivers' schedules yielding low, moderate and high fatigue levels as calculated by Dawson and Fletcher's (2001) fatigue model. The results indicated that fuel use, draft forces (train stretching) and braking errors were highest at moderate fatigue levels and lowest at high fatigue levels (Figures 9.3 and 9.4). This resulted in a counterintuitive finding, suggesting an improvement in braking and fuel efficiency as fatigue levels rose from moderate to high. If data analysis had stopped here, with just these performance measures, researchers and the rail organisation would have been left with the impression that fatigue results in more efficient driving!

However, upon examining some of the other performance measures further, extreme speed violations (>25% above the limit) and penalty brake applications were found to increase with rising fatigue levels.

These data indicated that from low to moderate fatigue levels, drivers were becoming less efficient, while from moderate to high fatigue levels, drivers were becoming more fuel efficient but less safe. Differences were clearest when drivers reached sections of track where they were required to slow the train down. In fact, what was happening is illustrated in Figure 9.5. In this example, the driver is on a descending section of railway, approaching a speed restricted section that requires a reduction in train speed of 20 km/h, a circumstance that typically requires drivers to apply the air brake. Drivers at a low level of fatigue applied the brake and minimised the speed violation. Drivers at a moderate level of fatigue applied the brake more heavily (making more braking errors). Drivers at a high

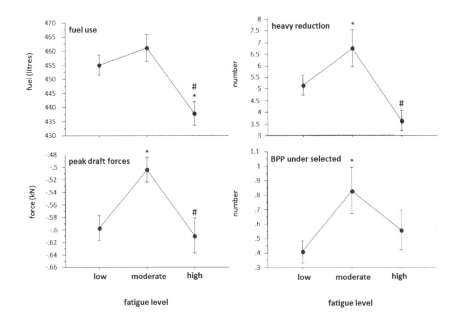

Figure 9.3   Simulator study example: Fuel use, heavy brake reductions, draft forces and brake pipe pressure (BPP) under-selected at low, moderate and high fatigue levels. * indicates a significant difference from low fatigue levels; # indicates a significant difference from moderate fatigue levels (p<0.05) (data from Dorrian, Roach et al., 2007)

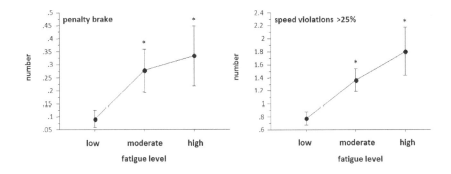

Figure 9.4   Simulator study example: Penalty brake applications and extreme speed violations at low, moderate and high fatigue levels. * indicates a significant difference from low fatigue levels (p<0.05) (data from Dorrian, Hussey et al., 2007)

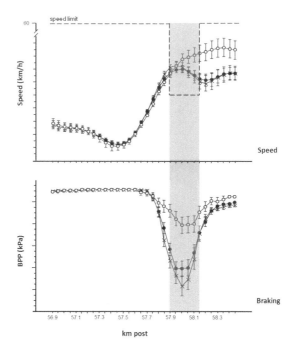

Figure 9.5   **Simulator study example: Brake use (upper panel) and corresponding train speed (lower panel) for drivers at low (filled circles), moderate (crosses) and high (open circles) fatigue levels during a speed limited section (from 60 to 40 km/h) on descending track grade. The dotted line represents the track speed limit (data from Dorrian et al., 2006).**

level of fatigue were much less likely to apply the brake at all (appearing more fuel efficient), resulting in an extreme speed violation.

    Thus, at high fatigue levels, errors involving a failure to act (often referred to as errors of omission) increased, whereas incorrect responses (often referred to as errors of commission) decreased (Reason, 1990). The differential effect of fatigue on error types can be explained through disengagement with the train at high fatigue levels whereby interaction with the train reduced dramatically and accident risk increased (for more detail see Dorrian, Roach et al., 2007). In other words, the change in performance from moderate to high levels of fatigue represented an efficiency/safety trade-off, as drivers became so fatigued they lost the mental energy to even attempt to control train speed appropriately (Figure 9.6). More information about the tensions in the efficiency/safety objectives in freight operations is discussed in the previous chapter (see 'Service delivery').

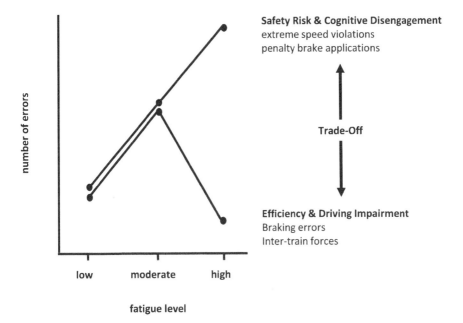

Figure 9.6   **Simulator study example: Schematic diagram showing the different patterns in errors relating to efficiency and impairment (e.g. braking errors) compared to those related to safety risk and disengagement with the train-driving task (e.g. speed violations, penalty brake applications) (adapted from Dorrian et al., 2007a)**

Interestingly, it should be noted that speed restrictions following moderate to heavy descending grades were more likely to be sensitive to fatigue. In contrast, performance was less affected during undulating territory, where a higher level of interaction was required to control the train (Dorrian et al., 2006). This illustrates the fact that it is not only performance measures that have differential sensitivity to fatigue, but the way the track is designed also influences the relationship between fatigue and performance, and should be kept in mind when designing track sections for testing in the simulator.

Thus, the results of the simulator study suggested that drivers at high levels of fatigue might disengage from the train-driving task, with potentially extreme safety consequences. The obvious question about these results is heavily related to the important issue of simulator fidelity, which was the focus of the previous chapter. Indeed, despite all the advances and improvements in simulator fidelity, a key issue with simulators is the inability to simulate consequences, which can result in a lack of motivation to perform optimally, as captured in the quote below:

> I discovered something else in the flight simulator. I sensed a shift as I became
> attuned to the simulator's pace and friction-free way of being ... worrying about
> real-life consequences fell away because they didn't matter any more. They'd
> been deleted. That was part of the freedom, I'm sure, but also part of the ennui
> that set in after so many highs and just as immaterial crashes. After a while, the
> game became rather rote and boring; nothing much mattered any more. (Quote
> about the experience of consequences in simulation from Menzies [2005])

Therefore, we wanted to know if the extreme safety impact of fatigue observed in the simulator happened because there were no real consequences for speeding, derailment or penalty brake applications in a simulated environment, and whether the concerns for safety in the real world would change the way fatigue affects driving performance. With this in mind, a follow-up study was conducted using data loggers in real trains, in order to measure the effects of fatigue in an environment where consequences were taken very seriously.

Comparing Simulator Results with Findings from Real Trains

In contrast to simulators, data loggers provide a record of driving behaviour in real trains, collecting information on braking, throttle use and fuel consumption at frequent intervals (e.g. one every second). Monitoring performance using data loggers is therefore less likely to have the issues with motivation that are possible in simulators, but the knowledge that you are being monitored is inescapable and may affect performance in the real train. The awareness of performance monitoring may, for example, stimulate the driver to put in the extra effort to drive well. In these circumstances, motivation effects are likely to be positive, making drivers more accountable for their behaviour and producing a safety improvement effect. However, where data loggers are integrated within standard company practice (some operators regularly download and review logger records), it can reduce acute motivation effects and result in a milder (more realistic and sustainable) general sense of performance awareness, accountability and ultimately, safety.

The following section will focus on the data logger follow-up study to the simulator study described above. Both studies were designed to investigate the effects of work-related fatigue on train-driving performance, and were therefore expected to generate roughly equivalent findings. However, at first glance, the findings did not appear consistent, and the synergy between the two studies revealed more about the research question than either of the studies alone.

Loggers were downloaded on 50 trains operated by pairs of male train drivers (24 to 56 years old) travelling on an Adelaide-Melbourne corridor (422 km, approximate trip time was 7.3 hours). As above, Dawson and Fletcher's (2001) fatigue model was used to calculate fatigue scores based on work history, and classified as low, moderate or high fatigue. Highly fatigued drivers used significantly more fuel ($p < 0.05$, see Figure 9.7) such that when compared to drivers with low fatigue, conservative estimates indicated these extra costs amounted to over \$3,500

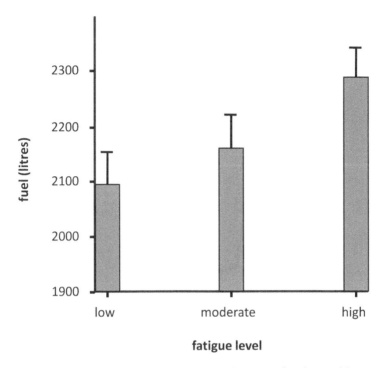

Figure 9.7 **Data logger study example: Fuel use (L) for the lead locomotive for drivers at low, moderate and high fatigue levels (data from Dorrian, Hussey et al., 2007)**

per week. The increase in fuel use was explained through decreases in dynamic brake and throttle use and increases in heavy brake violations (air brake applied too hard), highlighting that a more reactive driving style was being applied. An example of this is presented in Figure 9.8, where the speed limit on a section of track changes from 110 km/h to 95 km/h, and then back up to 110 km/h. The actual speed of the train is represented by the (solid) black profile line, with the (solid) grey line representing the preferred or optimal profile for this track (if the driver was driving to maximise the momentum of the train and conserve fuel by minimising the use of power and brake – we will revisit this concept of optimal driving with more of a critical eye later in this chapter in 'The Concept of Optimal Driving'). The train is naturally increasing in speed due to the downhill grade at that point in the track (km post 570 km) and exceeds the maximum track limit. The driver applies the air brake, which results in the train slowing down enough to comply with the reduced speed limit. However, since the brakes were applied too heavily, the speed continues to reduce to well below the speed limit (closer to 80 km/h) and the driver is required to use the throttle to increase the speed of the train back to an optimal level (95–100 km/h). This is inefficient and therefore costly in terms of fuel.

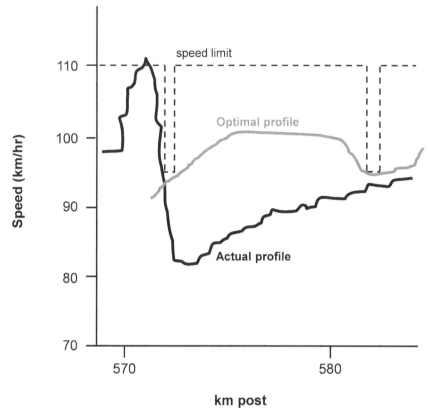

Figure 9.8   Data logger study example: Speed (km/h) during a track
section where the speed limit is changing from 110 to 95 km/h.
The solid black line indicates the actual speed for a driver at a
high fatigue level. The solid grey line indicates the preferred or
optimal speed profile for that section of the track.

Interestingly, in this study, no significant differences were found in terms
of average speed or speed violations. Consistent with the simulator results,
examination of track sections revealed significant differences in braking, fuel
use and speed, depending on track grade (for more detail see Dorrian, Hussey
et al., 2007). As mentioned earlier, since both the simulator and data logger
studies were designed to investigate the same research question, it was expected
that results would be equivalent. Initially however, this did not seem to be the
case. Specifically, at high levels of fatigue in the simulator study, extreme speed
violations increased and braking errors and fuel use decreased, whereas in the data
logger study, speed changes were not significant and fuel use increased. However,
the two studies illustrate the same phenomenon. That is, fatigue acts to reduce the

planning window, leading to a more reactive driving style characterised by late and/or incorrect brake use (including lack of brake use). The difference between the studies reflects a difference in the Safety/Efficiency Trade-Off. That is, in the simulator, with minimal consequences, the drivers became disengaged from the train-driving task and let the train run itself, resulting in extreme speed violations. In the data logger study, where there were real consequences to sub-optimal driving, safety was conserved at the expense of efficiency. This relationship is illustrated in Figure 9.9.

Given the lack of consequences in the simulator study described above, you may be thinking, why would we conduct simulator studies, why wouldn't we just do our research in the real world using technologies like the data loggers? The primary limitation of data loggers is the inability to control the conditions of each trip. As discussed above, duration, complexity and interest influence the way in which fatigue affects performance. It has thus been argued that the effects of fatigue on train driving are dependent on the characteristics of the driver, track, train and environment, which will determine duration, complexity and interest (Dorrian et al., 2006). For example, mechanical difficulties with the train or track or adverse weather conditions may make an ordinarily straightforward trip extremely complex. Similarly, it may be possible to choose particular types of trains and track sections for research, but it is impossible to control factors such

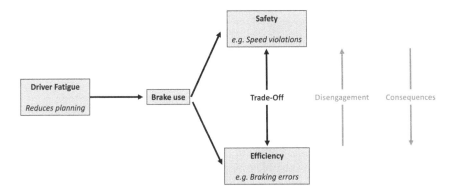

Figure 9.9   Combining the results of the simulator and data logger studies: Driver fatigue reduces planning for upcoming track features such as speed restrictions and affects the way drivers use the brake to try to slow the train. This can manifest in terms of reduced efficiency or safety. The lack of consequences in the simulator resulted in highly fatigued drivers who were more likely to become disengaged and therefore more likely to compromise safety. In contrast, in the real train, where consequences were critical, performance impairment was more likely to manifest in terms of reduced efficiency.

as weather, train cargo, loading or malfunctions. For these reasons, research using data loggers is likely to require large numbers of trips in order to reduce some of the 'noise' from such factors. These however, can be controlled in the simulator.

Arguably, therefore, the two methods can be used together to provide complementary views of the train-driving task under conditions of fatigue using environments with different levels of control and differences in possible motivation effects. In essence, these methods can be used to address the same research question, drawing from the benefits of both while minimising their limitations. Where simulators are used as the primary research tool, careful consideration should always be given to the possible perceptions around consequences and how these may influence driver motivation, engagement and performance. Moreover, where results are influenced by a lack of motivation, as in the example study described above, this does not mean that findings should be ignored, but rather that they should be interpreted in the appropriate context. Comparison with other studies and methodologies (such as data logger work) can be very illuminating. In industries analogous to rail, the problem has been minimised by integrating simulator measures within standard company practices. For example, in aviation, performance in the simulator is inherently tied to performance review (e.g. Flight Observation Quality Audit [FOQA], Helmreich, Klinect and Wilhelm, 2001) so pilots take simulator performance very seriously.

The way train simulators are used within industry can influence the way they are perceived by drivers and therefore levels of engagement and motivation to perform.

Taken together, the above studies show why it is important to consider the motivation to perform in the simulator, and why measures should be carefully selected in order to provide a complete picture of the way performance is influenced by a particular factor of interest. Using the fatigue example, the measures given in Table 9.3 can be classified according to whether they are best considered to be indicators of (1) general driving outcomes or driving efficiency, (2) the degree of performance impairment present or (3) safety risk or degree of disengagement with the driving task (Table 9.5). Studies should preferably include at least one measure from each of these categories, in order to provide a complete picture of the performance changes associated with increased fatigue (and thus adequately capture the Safety/Efficiency Trade-Off). The effects of other factors of interest could be considered in a similar fashion.

Table 9.5   Examples of train-driving measures that can be used to detect fatigue classified according to whether they best represent driving efficiency, impairment or safety risk and disengagement

Measure	Outcome or efficiency	Driving impairment	Safety risk and disengagement
Fuel use	X		
Fuel rate	X		
Brake use	X		
Throttle use	X		
Throttle changes	X		
Dynamic brake use	X		
Trip time	X		
Buff and draft forces	X		
Heavy brake reduction		X	
BPP over-utilised		X	
No split service		X	
Fast throttle changes		X	
Average speed			X
Maximum speed violation			X
Speed violations			X
Penalty brake applications			X
SPAD			X
Independent brake violation			X

The Concept of Optimal Driving

When conducting research, we tend to be more interested in relative changes (e.g. fatigue makes drivers use more fuel), but in operational settings, establishing benchmarks for optimal driving is a more frequent goal. How do we know whether an individual's driving is indicative of adequate training, route knowledge, experience, alertness, attention and so on? Similarly, how do we know whether an individual's driving style is catering for competing safety and performance goals? These questions centre on the concept of optimal driving. In the example of urban passenger operations, this would involve departing a station to arrive at a destination having met the schedule, remained safely within prescribed speed limits (and other movement authority), delivered a smooth passenger ride, and minimised energy consumption. Thus, under routine conditions, optimal driving is the assumption that any one journey can be performed on a trajectory that will effectively satisfy *all* service goals. In order to achieve optimal driving, the driver would require knowledge of the speed limits, the track's gradients and curvature (core route knowledge), the train's rolling resistance and tractive effort (traction knowledge) and beyond this, any momentary redundancies in the stopping pattern.

This is all very well but there are a number of key disturbances that must also be managed; environmental disturbances, for example those that impact railhead adhesion (see Figure 9.1) would change the way that route knowledge and traction knowledge are applied. Similarly, the basic human factors of alertness and attention would impact performance.

> **In performance testing it is not only useful to understand how to measure performance, but also to have an idea of what constitutes 'good'.**

The key message is that optimal driving is very dynamic and influenced by countless factors. For example, a driver who arrives at a station 10 min later than planned may still have done so optimally if the train has been driven through rain and fog. For this reason, benchmarks that constitute 'good' previous journeys, and which are also used as direct comparisons of subsequent performance, do not necessarily set the bar for what is an optimal drive. Instead, they reflect a journey with input variables that were perhaps easier to optimise, and by extension, happened to satisfy all measurable service delivery criteria. Beyond traditional driving practices, the international rail industry has adopted two main approaches for establishing optimal driving benchmarks, each of which would be measured differently. These can be broadly categorised as either instructional or adaptive modes of train driving, and must be considered carefully when measuring performance in operational settings.

Instructional train-driving modes are performed under a supervisory system (Naweed, Hockey and Clarke, 2009) that monitors and corrects speed transgressions (e.g. Automatic Train Protection). Some of these systems are communicated via in-cab displays, such as the European Train Control System (CENELEC, 2005) shown in Figure 9.10. In these interfaces, the perimeter of the speedometer is embedded with a series of movement authority zones that must be pursued or *tracked*, and very efficiently, in order to prevent the braking mechanism being triggered. Optimal driving based on instructional modes is almost entirely dictated by computer programming, thus performance testing would likely pay heed to measurable outcomes, safety risk and disengagement, and generally focus on speed. Examining performance based on speed violation thresholds (mild, moderate or extreme, see Table 9.3) and deviations from the movement zones may reveal much about the way that these systems are tracked, and enable effective diagnosis of performance decrement.

Adaptive approaches provide optimal driving benchmarks by taking into account the track and train characteristics to produce optimal speed, power and braking profiles for a particular drive. However, unlike instructional modes, there is no intervening supervisory system, and these parameters are recalculated to reflect throttle changes – thus the optimal pathway adapts to the decisions performed *by* the driver. Examples of these programs include the Metromiser for

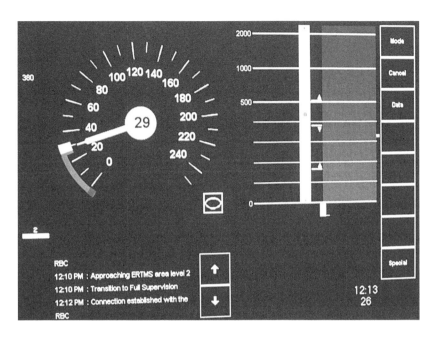

Figure 9.10   Picture of the European Train Control System (CENELEC, 2005) showing movement authority tracking zone in the speedometer

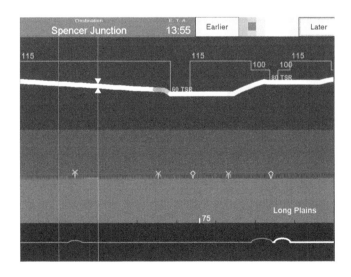

Figure 9.11   Screenshot of Freightmiser showing the track and train characteristics for optimal driving (Howlett and Pudney, 2000)

urban passenger trains, and the Freightmiser (shown in Figure 9.11) for long-haul trains (Howlett and Pudney, 2000; Pudney and Howlett, 1994). Such programs are based on principles to reduce fuel consumption by maximising train momentum, following a drive-coast-brake strategy, and they also provide in-cab displays for the driver that advise when to power, coast and brake in order to follow optimal energy use profiles. Investigations have indicated that these tools may be used to successfully improve on-time running and reduce energy consumption, though they may fall foul to the tendency to resist technology (see chapter 3).

For Metromiser, coasting time between stopping points (e.g. stations) is optimised based on required arrival times. For long-haul trains, the picture is a little more complex and so Freightmiser focuses on the concept of speed holding, taking into account forces on the train resulting from powering, braking, aerodynamic resistance and gradient. As such, the program uses track and train characteristics to calculate an optimal speed profile for the train. To follow this profile, the driver would drive to: (1) arrive on time (arriving early, or falling behind and trying to catch up wastes energy); (2) drive at a relatively constant speed, which is somewhere near 'hold-speed' or the average speed required to arrive on time; (3) avoid brake applications at higher speeds, coasting and slowing first; and (4) be aware of upcoming hills in order to increase speed prior to steep ascending grades and coast prior to descending grades (Howlett and Pudney, 2000; Pudney and Howlett, 1994). This approach requires a high degree of track knowledge and awareness, and forward planning.

Optimal profile technologies can be used to assess performance in operational and research contexts.

Figure 9.12 displays three actual versus optimal speed profiles from the data logger study. Section A is on an undulating track section with a series of speed restrictions, and shows that the driver is driving a lot closer to line speed (and exceeding the speed limit) than the optimal profile. The major deviation between the two profiles occurs near the 570 km post, where the actual speed profile dips dramatically as the result of an air brake application (73 to 65 psi). Section B is also a track section with numerous speed limit changes and an undulating track grade, but in this example, the severe dip in the speed profile is the result of a dynamic brake application. Sections A and B both illustrate examples of a failure to plan for speed restrictions, with late and heavy brake applications, resulting in dramatic slowing and the requirement to power and accelerate back to line speed. This is the pattern discussed earlier that we see in fatigued drivers. In contrast, Section C is an uphill section. In this example we can see that the speed profile of the driver matches the optimal profile.

Therefore, it can be seen how optimal profile technologies can be used to assess performance but it should be noted that such comparisons are not always straightforward. For example, it is critical to ensure that the track information

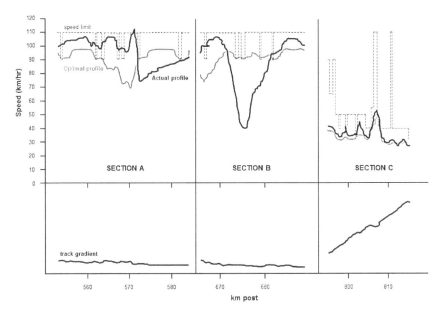

Figure 9.12 Data logger study: Three track sections (A,B,C) from the data logger study on freight trains on one Australian corridor. Speed limits (km/h) are represented by the dashed lines. The actual speed profile is represented by the black lines, the optimal speed profile by the grey lines. Track grade is shown in the lower panel.

used by the optimal profile program is completely accurate. Figure 9.13 shows deviations from the optimal profile that raise questions about the accuracy of track information. Specifically, at point a*, it could be questioned whether a temporary speed restriction is in effect, and at point b*, whether the speed restriction is correct. Clearly, these tools should be embedded in the context of complete and accurate information, and if this is not done thoroughly, it can be very easy to be misled by the output.

Selecting and Integrating Additional Measures of Performance

There are many additional measures of performance external to the simulator system that can be used to derive a more detailed picture of operator functional state. Researchers and trainers may also want to collect measurements associated with arousal, attention, vigilance, alertness, sleepiness, workload and situation awareness. A complete review of these measures is beyond the scope of the present chapter (though measures of situation awareness are discussed at length in chapter 10) but some common approaches, including behavioural, physiological and

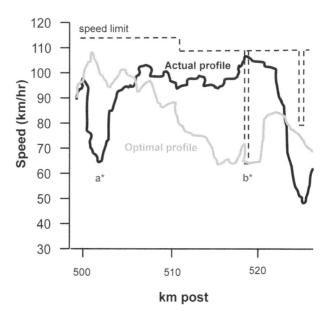

Figure 9.13   Data logger study: This track section includes mostly flat, slightly undulating grade. Speed limits (km/h) are represented by the dashed lines. The actual speed profile is represented by the black lines, the optimal speed profile by the grey lines.

self-report indicators of performance capacity, are discussed below. Much of the work in this area has been applied in the road transport domain but there has been some specific discussion in relation to trains. For in-depth reviews of driver alertness and fitness for work technologies, the reader is directed to Dinges, Mallis, Maislin and Powell (1998), Hartley, Horberry, Mabbott and Krueger (2000), and Horberry, Hartley, Krueger and Mabbott (2001); for a general discussion and on-road applications see Whitlock (2002); and in relation to the train cab and simulator see Oman and Liu (2007).

Behavioural Assays
As described in the introduction to this chapter, the task of driving a train involves memory (route knowledge), but it is also heavily reliant on monitoring the environment and planning for upcoming track features (Dorrian, Roach et al., 2007; Roth, 2000). In basic terms, the driver must be paying sufficient attention to their environment and searching for salient cues throughout the drive. The basic capacity to monitor and sustain attention is captured by many behavioural assays used in the performance-testing field for more than a century. A large number of these are stimulus-response type tasks, which require participants to attend

to a computer screen, a response box or an auditory source for a period of time (anywhere from 90 s to 20 min). A response is usually made by pressing a button whenever a stimulus appears, usually in the form of a light or noise, and the time to respond is subsequently recorded (i.e. time taken between the appearance of the stimulus and the button press). Stimuli typically occur at random intervals throughout the testing period, and an individual's response times over that testing period are considered to be an indicator of their ability to sustain attention over time (Dorrian, Rogers and Dinges, 2005).

Stimulus-response tasks abound with examples including: the Wilkinson Auditory Vigilance Task (e.g. Glenville, Broughton, Wing and Wilkinson, 1978), the Mackworth Clock (e.g. Williamson, Feyer, Mattick, Friswella and Finlay-Brown, 2001), four-choice vigilance (e.g. Lieberman, Tharion, Shukitt-Hale, Speckman and Tulley, 2002), visual vigilance (e.g. Magill et al., 2003), and the Psychomotor Vigilance Task (PVT, e.g. Dinges and Powell, 1985). To explore one of these in more detail, the PVT has been used widely, particularly in relation to sustained attention testing during sleep loss and fatigue (Dinges and Powell, 1985; Dorrian et al., 2005). This task is delivered in various formats, including desktop computer, response box (Figure 9.14, left panel) and via palm pilot (Figure 9.14, right panel). Participants watch the response screen and press a response button with their dominant hand when they see the stimulus appear. The task typically lasts for 10 min (Dorrian et al., 2005), however, 20-min (Ratcliff and Van Dongen, 2011), 5-min (Loh, Lamond, Dorrian, Roach and Dawson, 2004) and even 90-s (Roach, Dawson and Lamond, 2006) versions have been used. Given the time

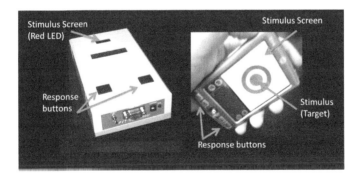

Figure 9.14 **Psychomotor vigilance task (PVT) depicted in response box (left panel) and palm pilot (right panel) formats. In the response box delivery, the stimulus is in the form of a red millisecond counter on the LED screen that begins at zero and stops when the button is pressed, displaying response time in milliseconds. For the palm pilot version, the stimulus is in the form of a target, and when the response button is pressed, response time in milliseconds is displayed at the centre of the target.**

constraints in an operational environment, 5-min versions, administered via palm pilot, are being increasingly employed to test sustained attention in the workplace (e.g. Ferguson, Paech, Dorrian, Roach and Jay, 2011; Ferguson et al., 2010).

In addition to the stimulus-response type assays, tracking type assays have also been used (e.g. Lamond and Dawson, 1999; Magill et al., 2003; Williamson et al., 2001). These types of tasks require participants to keep a cursor on a moving target with a random trajectory. An example is displayed in Figure 9.15. An example of a tracking task that has been used in industry studies is the Occupational Safety Performance Assessment Test (OSPAT, Romteck, Perth, 1998). While tracking tasks are useful for measuring performance, it is important to note that their validity as indicators of complex tasks like train driving has been questioned (Hartley et al., 2000). It should also be noted that these tasks increase the overall duration of the testing session, and depending on test and sensitivity requirements, add anywhere from 90 s to 20 min to the time spent on the simulator. For this reason they add substantially to experimenter and participant burden. Further, since such tasks are also likely to show aptitude effects (i.e. different people can perform at different levels), they are used most effectively in a repeated measures context, where individuals are measured multiple times over. Changes relative to their own baseline (or 'typical') level of performance can therefore be assessed. Note however that the more a participant performs, the better they get, thus different tasks are also likely to produce learning or practice effects, and allowing for practice time before a data collection period may be important. For the PVT described above, there is a one to three trial learning curve for the 10-min version (Dorrian et al., 2005), which means that it is best for participants to practice it three

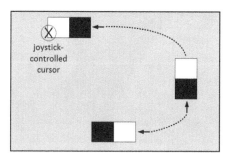

Figure 9.15   Example   tracking   task   (Information   Processing   and Performance Test System – Feyer, Williamson and Rassack, 1992). A 12 x 12 mm square following a random, winding path was presented on the computer monitor. Using a joystick to control a cursor, participants were required to keep the cursor on the square as much as possible. The test lasted 3 min. The score was given in terms of the percentage of the time that participants kept the cursor on the square.

times before a baseline is established. Finally, given that these are supplementary tasks woven into simulator use, processes for data capture, backing up and storage will add to experimenter burden.

With all of those issues in mind, you may well question why you would want to include behavioural testing in the first place. There are a number of advantages to collecting a basic indicator of performance capacity. As an example, we can again go back to the simulator study first examined in 'The Effects of Fatigue on Simulated Driving Performance' above.

As discussed earlier, the initial results of the study seemed to indicate, to our surprise, that at high levels of fatigue, drivers used less brake and less fuel, compared to low and moderate fatigue levels (Figure 9.16, left panel). This caused us first to wonder about whether our classification of fatigue levels was accurate – perhaps we were not looking at the data correctly? Fortunately, each time a driver had completed a run in the simulator, they had also completed a 10-min response-box version of the PVT. Data are shown in Figure 9.16 in the right panel and indicate clearly that our classification of fatigue levels was likely to be accurate, given that this test showed clear results in the expected direction (response times increased with increasing fatigue). This spurred us to investigate other performance metrics in more detail, and added weight to our final interpretation of study results, that with increasing fatigue, drivers drove more efficiently but with a serious trade-off in terms of safety (Figures 9.6 and 9.9). Indeed, basic tests of performance such as the PVT are frequently used in applied studies where they can be employed as a 'methodology check'. We have a great deal of literature on how PVT performance should change depending on the study methodology, and when we see the expected change, we can be confident that our experimental manipulation had the desired effect (Dorrian, Lamond, Kozuchowski and Dawson, 2008; Dorrian, Roach et al., 2007).

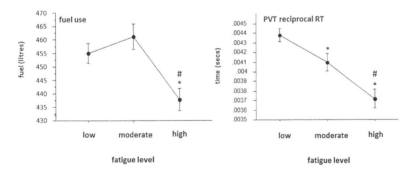

Figure 9.16   Simulator study example: Fuel use and PVT performance (reciprocal transformation of response times such that lower values indicate worse performance) at low, moderate and high fatigue levels. * indicates a significant difference from low fatigue levels (p<0.05) (data from Dorrian, Roach et al., 2007)

Another benefit of collecting PVT data in the simulator study was that we could use it as a general indicator of how long drivers' response times were likely to be at given fatigue levels. In particular, we were interested in long response times (between 0.5 s and 10 s). This allowed us to think in more detail about the effects that fatigue may be having on the drivers. For example, long response times became more frequent with increasing fatigue, and 10 per cent of highly fatigued drivers experienced lapses between 8–10 s in length. In practical terms, if the train was travelling at a speed of 50 km/h and a driver experienced a lapse of this duration, the train could travel nearly 140 m in this lapse time, far enough for serious operational consequences (Dorrian et al., 2006).

Secondary Tasks

Secondary tasks adopt stimulus-response and tracking type assays, but unlike the behavioural assays described in the previous section, they are carried out at the same time as the primary task. Thus, they are designed to temporarily divert attention away from the primary task, and the time taken to do so reflects the budget of the cognitive demand or attention that was allocated to the primary task in that moment. In this way, secondary tasks are generally described as measures of workload, spare attentional or visual capacity. Secondary task techniques have been shown to effectively assess cognitive demands in rail and other domains, both in the form of tracking type tasks (Askey and Sheridan, 1996) and stimulus-response paradigms (Hockey, Healey, Crawshaw, Wastell and Sauer, 2003). Secondary tasks can provide useful information, but as they are secondary in the overall context of the drive, they are not a pure indicator of attention and must be interpreted in that context. For example, very long responses could indicate that attention is being sacrificed in an effort to protect train driving, but it does not tell us where it is being allocated, and in what capacity, so the context of the primary task during the response is generally used to infer meaning.

> **System response times of vigilance-type systems have the benefit that they are collected while driving and therefore do not contribute to increased participant burden in terms of testing time.**

Given that many secondary tasks follow the stimulus-response convention, it is also useful to think about the relationship between these tasks and the in-built stimulus-response vigilance and station protection systems that are used in most trains. Such systems typically require the driver to push a button in response to a timed alarm, and where this response is not given, the emergency brakes bring the train to a stop (reviewed briefly in Dorrian et al., 2008). Thus, the importance of having the ability to respond in a timely fashion to a stimulus is explicitly acknowledged in the rail environment as a useful indicator of the basic capacity

   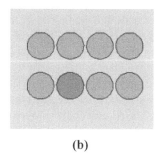

(a)   (b)

Figure 9.17   **(a) Picture of a stimulus-response secondary task response box used during research with a train-driving Microworld. (b) The eight buttons depicting circular icons correspond to the layout in the simulator. Randomly selected targets are cancelled by the corresponding response key.**

to function safely. In many simulators, the time taken to respond to the vigilance systems can be recorded and outputted for review.

Research simulators have drawn on the dynamics of the stimulus-perception-action relationship already found in the driver cab to engineer novel secondary tasks. Figure 9.17 shows the secondary task targets in a train-driving Microworld (i.e. a computer-based abstract representation of the train-driving environment, see 'The Microworld' in the previous chapter) and the corresponding response keys on an input device (Naweed et al., 2009). The secondary task in this simulator was made up of two rows of eight targets and occurred at random intervals. An active target was detectable through a change in contrast (second target on bottom row in Figure 9.17) and much like in-cab vigilance/protection devices, an auditory alert was used to provide the original frame of reference. In this particular task, the distinction of multiple targets with coordinated responses was implemented so that drivers would exert more effort to scan and cancel the task, and in doing so demonstrate more sensitivity in their responses to the task over critical driving periods.

Physiological Indicators
There are numerous technologies that can be employed to provide a fundamental indicator of physiological arousal. Most of these have focused on measuring brainwaves (Åkerstedt and Gillberg, 1990; Lal and Craig, 2001; Torsvall and

Åkerstedt, 1987; Wright and McGown, 2001) or eye-responses (Cabon et al., 1993; Torsvall and Åkerstedt, 1987; Wright and McGown, 2001). Other, perhaps more controversial technologies have measured facial tone, head nodding, body posture (reviewed in Hartley et al., 2000; Wright, Stone, Horberry and Reed, 2007) or electrodermal activity (reviewed briefly in Dorrian et al., 2008). Frequency and amplitude of brainwave activity and patterns in eye movements can be used for real-time monitoring of alertness while drivers are in the cab (Torsvall and Åkerstedt, 1987). This requires electrodes to be attached on the head of the driver, and the signals recorded and later scored by experimenters. Brainwaves and eye movements can be analysed in order to indicate periods of low arousal, drowsiness and dozing, and can therefore provide really interesting information about why performance recorded in the simulator changed in the ways that were observed. This is especially powerful when these data are synchronised with the performance data recorded by the simulator. As a broad example, if a driver was recorded to have missed a signal or failed to respond to the vigilance system, such extra information could shed light on whether the driver was asleep, dozing off or simply not paying attention. The disadvantage of such systems is that they are relatively costly, can be a little uncomfortable for drivers and take a trained experimenter time to score (i.e. turn the data from zigzagged lines to specific information about drowsiness).

Other types of eye measurement include glance data (Sodhi, Reimer and Llamazares, 2002), eye movements (Cabon et al., 1993; Torsvall and Åkerstedt, 1987; Wright and McGown, 2001), blinking (Verwey and Zaidel, 2000) and pupil response (Horberry et al., 2001). Some can monitor in real time during a drive using cameras fitted in the cab (Smith, Shah and da Vitoria Lobo, 2000) or via technology embedded glasses frames (Stephan et al., 2006), while others require separate testing time (Lamond, Dorrian, Kozuchowski, Hussey and Dawson, 2005). Clearly, it is very useful to know where a train driver is looking and for how long, but like the other psychophysiological measures, there are no free lunches. They add varying levels of participant burden, and again, take a trained experimenter time to score. They also vary in terms of the amount of prior testing/validation that has been done, their cost, ease of use (some can be very tricky to configure), simplicity of extraction and interpretation of data. These are important points to consider during the cost-benefit evaluation for an experiment. Some examples of physiological indicators based on ocular function include PERCLOS (Dinges et al., 1998), Eyecheck (Lamond et al., 2005) and OPTALERT (Stephan et al., 2006).

Other measures that can be of benefit in the simulator cab include audio and video recording. Vocal quality and word generation can be used to indicate fatigue (Bagnall, Dorrian and Fletcher, 2011; Harrison and Horne, 1997), oxygen deprivation (Cymerman et al., 2002; Lieberman, Protopapas and Kanki, 1995), disease (Yuceturk, Yilmaz, Eqrilmez and Karaca, 2002) and mood (Kleitman, 1963). Vocal recordings of 'expert narratives' or commentaries can also be very informative for review and research; asking drivers to 'think aloud' and explain how they are driving as they go has been used to capture situation awareness and decision-making processes (Hughes and Parkes, 2003). Audio and video records can facilitate the analysis of two-driver interaction, communication with control or other train drivers, body language and movement (e.g. changing positions to combat discomfort or sleepiness), facial expression, head nodding and long eye closures (indicative of extreme drowsiness) and/or response times to various stimuli in the cab (e.g. vigilance or station protection systems). Cameras can also be used to monitor where a driver is looking, though infrared equipment will likely be needed for trips at night.

Self-Report Measures

Subjective scales can provide a fast and effective way of recording perceptions of various aspects of functioning, such as fatigue, alertness, workload, situation awareness and the ability to perform different tasks. These scales can take different formats, and Figure 9.18 illustrates four different self-report methods for asking about fatigue. The top panel shows a Visual Analogue Scale (VAS), which is typically a 100-mm line anchored with semantic opposites that participants place a mark on (shown in red). VAS are simple to explain and easy to apply (Bond, Shine and Bruce, 1995), and are scored by measuring where the participant mark

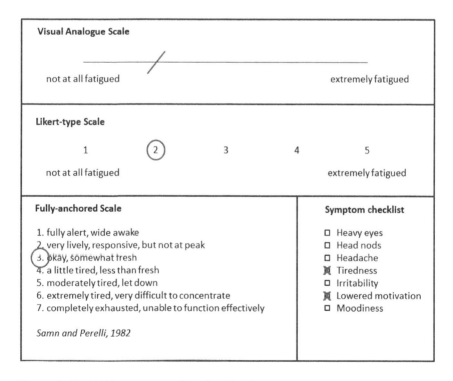

Figure 9.18 Different examples of subjective methods for measuring fatigue

sits in the line. The resulting score is out of 100 mm, which makes them relatively sensitive compared to some of the other methods (which may yield a number out of five, or seven, for example). The middle panel in Figure 9.18 gives an example of a Likert-type scale, which is a forced-choice scale. In the pictured example, the anchors are at extreme ends of the scale only, which is often done for simplicity, but they are also used with anchors for all responses, or with a 'neutral' anchor at the mid-way point. An example anchor for the neutral point on this scale may be 'neither alert, nor fatigued'. Likert scales are seen extremely frequently in social sciences research, and though they are generally not as sensitive as VAS, they may be given detailed anchors to provide a more specific response. In addition, responses are often faster to process, as they do not require measurement. The Likert scale in Figure 9.18 is an example of a fully anchored scale, the Samn-Perelli Fatigue Scale (Samn and Perelli, 1982), which is found in abundance, well-validated and widely used. Some have descriptive anchors, while some are completely behavioural. Finally, at the bottom right of Figure 9.18 is an example of a symptom checklist, which requires identification of one or more descriptors from a series. When choosing a self-report scale, consideration must be given to what the data from the scale will be used for, and how it will be analysed. In this example, the data may be analysed in terms of a score of 0–100 (VAS), 1–5 (Likert), 1–7 (Samn-Perelli) or number of symptoms (checklist). The Likert scale could be turned into a binary outcome indicating fatigue (i.e. above neutral) versus indicating no fatigue (i.e. below neutral). Data from the checklist could be treated separately for each symptom, such as heavy eyes (yes/no), and headache (yes/no). It all depends on the particular aims of the investigation.

Going beyond the individual scale level, various questionnaires, inventories and omnibus instruments (single instruments containing multiple measures) have been designed to provide a more comprehensive measure of different constructs. These tend to be made up of a series of scales and can take anywhere from 1 min to over 30 min to complete. For example, the NASA Task Load Index (NASA-TLX) takes 5–10 min to complete, including six scales, and gives information regarding overall workload based on mental demand, physical demand, time demand, performance, effort and frustration (Hart and Staveland, 1988). The profile of mood states (POMS) contains a series of 65 adjectives, which participants must respond to on a series of Likert scales indicating how much this describes their current mood. This instrument takes 10–15 min to complete and provides summary indicators of total mood disturbance, tension-anxiety, anger-hostility, fatigue-inertia, depression-dejection, fatigue-activity and confusion-bewilderment (McNair, Norr and Droppleman, 1971). One type of measure of particular salience to the train-driving environment is situation awareness, which has been investigated using both behavioural and self-report techniques. For a full discussion of situation awareness measurement, the reader is directed to chapter 10.

It should be noted that self-report techniques are necessarily collecting individuals' reported responses to particular questions. They do not always correspond with objective measurements of similar constructs, or even to how the individual really feels about something. They can be influenced by many factors, including workload (Dinges, 1990), mood, boredom, hunger (Dinges and Graeber, 1989), social desirability (i.e. what the person perceives as the socially desirable way to respond) (Moller and von Zerssen, 1995), how long they have been performing the task under investigation, time-of-day, context and motivation to perform. Nevertheless, if the context is taken into consideration, they can provide valuable additional information. Lastly, whilst self-report scales are typically given in paper and pencil format, it may be more time-efficient to use palm-pilot or laptop devices to collect data, saving time on data entry. Needless to say, all of the measures discussed in this section have strengths and weaknesses for the pursuit of train simulator-based research. Table 9.6 summarises the advantages and disadvantages of behavioural assays, physiological indicators and self-report measures.

Table 9.6   Summary of advantages and disadvantages of the measures discussed above

Measure	Advantages	Disadvantages
Behavioural assays	• Simple indicators of functional capacity • Stimulus-response tasks have clear parallels with vigilance safety systems in the rail cab. • Can be used as a 'methodological check' for known effects • Can give additional context to train-driving performance measures • Some can be purchased for minimal cost (e.g. palm-pilot version of PVT).	• Validity as indicators of complex operational performance has been questioned • Add to participant time burden • Require practice and baseline assessment • Add to data output and processing requirements (although most have simple programs to help to analyse the data) • Some versions are more expensive to purchase.
Physiological indicators	• Can provide basic indicators of physiological arousal • Can often be used in a real-time monitoring situation, where data can be time-locked to driving performance in the simulator • Can provide extra information about why performance changes occurred	• Tend to have associated expense • EEG/EOG require trained operators to process data • Different measures have been subjected to differing amounts of validation studies • Add to participant discomfort burden

Self-report measures	• Can provide information about individual concepts (scales) or whole constructs (questionnaires, inventories) • Low cost • Individual scales tend to be fast.	• Questionnaires can add to participant time burden. • Subjective nature must be acknowledged and responses can be influenced by many factors.

Additional Measures: How Will I Choose?

Many considerations will invariably factor in your decision to include one or more of the additional measures into your simulator system. It is important to know what information a particular measure will add, and how it would help you to interpret the data that you collect. You also need to know about the validity of the measure, how widely it has been used, and what others think about it. Equally important is its cost and how much it burdens simulator users in terms of time and discomfort. Lastly, what can you do with the data, and how easy is it to extract? These questions go hand in hand with the issues of scenario design, data acquisition and overall research tractability discussed at the end of the previous chapter. In short, there are likely to be many questions around the tractability of the extra measure and the value it adds to your research. Some of these questions are summarised in Table 9.7, and form an excellent start point with which to get to know these measures.

Table 9.7   Questions to ask when considering extra simulator measures

Information gain	• What information will it add? • Why do I want to know this information? • Will it help me interpret the performance data I collect from the simulator?
Validation	• Has it been validated? How extensively? • Am I convinced by the validation studies? Who conducted the studies? Did they disclose any conflict of interest (financial or otherwise) that may cause me to question their results? How reputable were the forums in which the studies were published? Were they considered to be primary, peer-reviewed studies? • How widely has the device been used? • What do other people say about it?
Cost and burden	• How much does it cost? • How much technological complexity does it add (maintenance, repair etc.)? • How much burden does it place on participants (time, discomfort)?
Output	• How easy is it to harvest, store and interpret the data from it? • Can I synchronise the data to the simulator measurements of performance?

The Simulator System OUTPUT: Data Sampling and Format

The last thing to consider is the OUTPUT stage of the simulator system and the format of the data, specifically: (a) frequency and type of data sampling, and (b) required output format. For training and review purposes, a lower data-sampling rate provides a sufficient detail of performance, while more detail is generally required for research. Performance review typically involves examining the performance of an individual across time, where programs sample it every few seconds or so to provide a 'snapshot' at each time point. Figure 9.19 displays an example screenshot from a viewing program outputted from a session in a freight train simulator. The screen shot shows a variety of outcome or efficiency performance measures (draft and buff forces, acceleration, throttle position, automatic brake position, time) and indicators of safety risk and disengagement (speed, horn application, speeding infractions). It also shows topographical track elements (track height, radius, track features) providing reference points with which to diagnose performance. The viewing program captures a detailed overview of the driving session and allows the session to be 'eye-balled'. For industry, a key benefit of these programs is that they transpose performance in an altogether readable and familiar format. We can see from the speed data in Figure 9.19 that the driver did not stop the train at any point during the session, and that the automatic brake was not applied. Speed and acceleration were therefore entirely regulated with the throttle. We can also see that the buff and draft forces alternated as the train climbed and descended the gradients.

Although the viewing program displayed in Figure 9.19 has sampled performance at a high rate (second-by-second), and provided a range of measures, more would generally be needed for the empirical research context. Research often aims to investigate the performance of sample populations, perhaps distinguished by group differences such as fatigue level (Dorrian, Roach et al., 2007; Dorrian et al., 2006), or shift design (Thomas and Raslear, 1997). Using interval (or time-stamped) data to investigate groups of drivers (derived from the position of the lead car in Figure 9.19) can be problematic because drivers will not all travel at the same speed across different sections of track. Moreover, such differences will likely be exacerbated by variables such as fatigue and shift design. Thus, for programs with interval data collection, drivers will be at different track locations for each time point recording or 'snapshot'. Therefore, in order to line up the data across drivers for analysis, these data need to be converted from an interval-stamp to a distance-stamp base, which is often difficult to do, and can be imprecise and very time-consuming. As an example, in the study described in 'The Effects of Fatigue on Simulated Driving Performance' data were collected in 4.5-s 'snapshots'. Figure 9.20 displays an example of raw data from a single driver in 4.5-s time intervals. Whilst this did not present a problem when data across a whole trip was being investigated (Dorrian, Roach et al., 2007), it was a bigger issue when examining individual track segments (speed restricted sections), where it became necessary to specifically distance-lock the data across drivers.

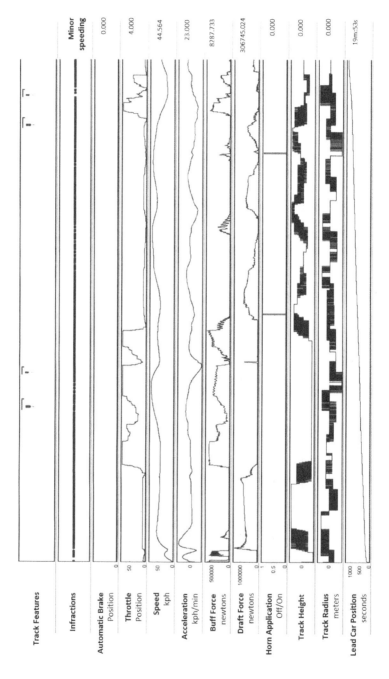

Figure 9.19   Example of a generic post-session viewing program

Time	Force	Kilometre	Metre	Grade	Speed	Acceleration	THTLDB	Amps	LeadBPP	LeadBCP	RearBPP	DraftSS	DraftPK	BuffSS	BuffPK	EventBreak
00:00.0	115392	384	0	0.08	0.1	-4.97	0	0	496	20	496	0	0	0	0	0
00:04.5	115392	384	0	0.08	0	0	0	0	496	344	496	57	0	0	0	0
00:09.0	115392	384	0	0.08	0	0	0	0	496	344	496	57	0	0	0	0
00:13.5	115392	384	0	0.08	0	0	0	0	496	344	496	57	0	0	0	0
00:18.0	115392	384	0	0.08	0	0	0	0	496	344	496	57	0	0	0	0
00:22.5	115392	384	0	0.08	0	0	3	77	496	344	496	57	0	0	0	0
00:27.0	115392	384	0	0.08	0	0	0	93	496	344	496	57	0	0	0	0
00:31.5	115392	384	0	0.08	0	0	0	0	496	344	496	57	0	0	0	0
00:36.0	115392	384	0	0.08	0	0	3	120	496	330	496	57	0	0	0	0
00:40.5	115392	384	0	0.08	0	0	3	268	496	110	496	57	0	0	0	0
00:45.0	115394	384	0	0.08	0.8	13.99	3	301	496	0	496	124	0	0	0	0
00:49.5	115399	384	1	0.08	2.2	22.54	4	380	496	6	496	155	0	0	0	0
00:54.0	115411	384	4	0.08	4.1	26.43	5	412	496	6	496	182	0	0	0	0
00:58.5	115431	384	10	0.08	6.4	36.81	8	498	496	6	496	213	0	0	0	0
01:03.0	115462	384	19	0.04	9.1	41.53	8	413	496	6	496	249	0	0	0	0
01:07.5	115503	384	32	0.04	11.6	39.1	8	486	496	6	496	235	0	0	0	0
01:12.0	115555	384	47	0	13.9	31.76	8	427	496	6	496	204	0	0	0	0
01:16.5	115617	384	65	0	16.3	30.26	8	413	496	6	496	191	0	0	0	0
01:21.0	115687	384	87	0	18.6	29.75	8	401	496	6	496	182	0	0	0	0
01:25.5	115767	384	111	0	20.7	27.05	8	377	496	6	496	169	0	0	0	0
01:30.0	115855	384	137	0	22.5	23.55	8	342	496	6	496	146	0	0	0	0
01:34.5	115950	384	166	0	24.2	22.01	8	326	496	6	496	137	0	0	0	0
01:39.0	116053	384	197	0	25.8	21.26	8	313	496	6	496	128	0	0	0	0
01:43.5	116161	384	230	0	27.4	20.61	8	300	496	6	496	120	0	0	0	0
01:48.0	116276	384	265	0	28.9	20.01	8	290	496	6	496	115	0	0	0	0
01:52.5	116397	384	302	0	30.4	19.67	8	280	496	6	496	111	0	0	0	0
01:57.0	116525	384	340	0	31.9	19.01	8	272	496	6	496	102	0	0	0	0
02:01.5	116657	384	380	0	33.3	18.36	8	264	496	6	496	102	0	0	0	0
02:06.0	116796	384	422	0.33	34.6	16.88	8	258	496	6	496	93	0	0	0	0
02:10.5	116940	384	466	0.83	35.7	13.71	8	252	496	6	496	88	0	0	0	0

Figure 9.20 **Example of raw data from one driver, time-locked in 4.5-s intervals**

Figure 9.21 displays the data for the same driver at the same point in the track for just two variables – brake pipe pressure in the lead loco and speed. These data are averaged over 100-m intervals. In the study, we used a distance interval of 50 m. From Figure 9.5 (on p. 222), it appears as though drivers applied the brakes at the same point prior to the speed restriction, regardless of their fatigue level. However, when converting the interval data to a distance-locked base, the highest resolution that could be achieved was 50-m increments (the distance a train would travel at a speed of 40 km/h in 4.5 s). Therefore, drivers could have a lapse in attention of up to 4.5 s and it would not clearly show in the braking time. Higher data resolution would certainly have been beneficial.

Therefore, data output requirements for training, review and research may be competing. That is, using data for training or performance review purposes may involve immediate examination of data for one individual using a specifically designed viewing program. Research typically requires a basic output format (e.g. ASCII) to allow data to be added to databases of multiple individuals

driver	run	variable	d384	d385	d386	d387	d388	d389	d390
1	1	Brake	496	496	496	496	496	496	475
1	1	Speed	61.35	63.75	65.90	66.70	66.00	63.70	59.70

Figure 9.21 **Example of raw data from one driver, distance-locked in 100-m intervals**

for processing and analysis in different programs (e.g. Access, Excel, SPSS, STATA, SAS). Therefore, if output is to be effectively used for training, review and research, the output options need to be general enough to suit a review/ monitoring program or an analysis program. As mentioned above, basic formats such as ASCII tend to be appropriate in most circumstances. To reiterate a key point in the previous chapter, specifying the way that data should look and how it should be outputted is an important but unfortunately, easily overlooked stage in procurement. At first glance, the range of data shown in Figure 9.19 appears comprehensive, and given that it captures performance by the second, it certainly has high resolution. However, the distance of the track is not shown, and if this were not available, conversion to distance-locking would be very difficult. Similarly, there is no indication of the speed envelope, brake pipe pressure, instances of the vigilance trigger (including time to respond) and the fuel running total, all of which would be critical for investigating and inferring performance from a research perspective.

> **The data output requirements for training, review and research may be competing.**

Conclusions

This chapter has described the simulator as an integrated system, and broadly shown how you may wield one to enhance performance (productivity, efficiency) and safety. The inputs of the system are the demands of training, performance review and research. We have the opportunity to upgrade existing technology to improve interoperability, and design new technology that is effective both for industry and research. For an arguably minimal cost, there could be substantial return on investment in order that both the review and research processes contribute to performance and safety in the workplace. In order to achieve this, industry and researchers must work closely together to lessen the focus on retrospective evaluation based on pre-existing knowledge and increase the focus on generating new knowledge. That is to say, training or performance reviews are used primarily to evaluate existing practice. A research dimension can allow for new ideas that can feed back into training and review to highlight strengths and weaknesses of current practice and develop more sensitive, effective or targeted practice.

At the processing level of the system, consideration for the most appropriate measures to provide a complete picture of performance is paramount. Choosing output measures to represent efficiency, impairment and safety is key, as is considering the unique environment of the simulator, especially in terms of the perceived consequences of poor performance in interpreting results. Converging evidence from other studies in the laboratory and the field, using different performance monitoring technologies (e.g. data loggers), can be useful.

Finally, in order to satisfy the requirements of training, performance review and research, the data output from performance-monitoring technologies should be:

1. in a suitable format for performance monitoring/review programs and data analysis programs (e.g. ASCII)
2. able to be manipulated to investigate or monitor the performance of individuals over time (interval data) as well as investigating group performance (distance-locked data); Global Positioning Systems for Data Loggers are likely to be extremely useful in this regard, however reliability and accuracy should be established.
3. accurate to an appropriate level of precision: This may need to be more detailed for research purposes, but as a rule of thumb, the sampling interval should be no longer than the time it would take a train to travel the maximum acceptable distance increment at the average speed of that track. For example, if the average speed was 40 km/h, a sampling interval of 4.5 s would allow a precision of 50 m, whereas a sampling interval of 2 s would allow a precision of 22 m.

Armed with current understanding of train-driving performance changes in fatigued individuals, and with ever-increasing technological advances, designing an integrated performance-monitoring system is timely and important. This system should be integrated in standard industry practice, ideally with the appropriate measures and output to inform training, review and research. Development of such an integrated system will provide a comprehensive understanding of the effects of different operating conditions on the train driver, and support better performance and safety management in the rail industry. Taken together, our two 'evaluating your train simulator' chapters have discussed a wide range of ideas and points around the physical and task environments of your current or future simulator.

Acknowledgements

For their intellectual, logistic and financial contribution to the studies described in this chapter, the authors would like to thank The Australian Rail Consortium Shiftwork and Workload Study, Frank Hussey, Pat Wilson, Gregory Roach and Drew Dawson (Appleton Institute, University of Central Queensland), Peter Pudney (University of South Australia) and the participating rail operators and study volunteers.

References

Åkerstedt, T., and Gillberg, M. (1990). Subjective and objective sleepiness in the active individual. *International Journal of Neuroscience, 52*, 29–37.

Åkerstedt, T., Torsvall, L., and Froberg, J.E. (1983). A questionnaire study of sleep/wake disturbances and irregular work hours. *Sleep Research, 12*, 358.

Askey, S., and Sheridan, T.B. (1996). *Safety of high-speed ground transportation systems* (DOT-VNTSC-FRA-04-06). Cambridge, MA: Department of Transportation.

Bagnall, A.D., Dorrian, J., and Fletcher, A. (2011). Some vocal consequences of sleep deprivation and the possibility of "fatigue proofing" the voice with voicecraft® voice training. *Journal of Voice, 25*(4), 447–61.

Bond, A., Shine, P., and Bruce, M. (1995). Validation of visual analogue scales in anxiety. *International Journal of Methods in Psychiatric Research, 5*, 1–9.

Cabon, P., Coblentz, A., Mollard, R., and Fouillot, J. P. (1993). Human vigilance in railway and long-haul flight operation. *Ergonomics, 36*(9), 1019–33.

CENELEC. (2005). European Standard – Railway applications – Communication, signalling and processing systems – European Rail Traffic Management System – Driver-Machine Interface – Part 4: Data entry for the ERTMS/ETCS/ GSM-R systems *Comité Européen de Normalisation Electrotechnique* (Vol. CLC/TS 50459-4:2005).

Cymerman, A., Lieberman, P., Hochstadt, J., Rock, P.B., Butterfield, G.E., and Morre, L.G. (2002). Speech motor control and acute mountain sickness. *Aviation, Space, and Environmental Medicine, 73*(8), 766–72.

Dawson, D., and Fletcher, A. (2001). A quantitative model of work-related fatigue: background and definition. *Ergonomics, 44*(2), 144–63.

Dinges, D.F. (1990). *The nature of subtle fatigue effects in long-haul crews.* Paper presented at the 43rd International Air Safety Seminar, Flight Safety Foundation, 19–22 November 1990 Rome, Italy.

Dinges, D.F., and Graeber, R.C. (1989). *Crew fatigue monitoring.* Paper presented at the Proceedings of the Second Regional Workshop on Crew Performance Monitoring and Training, Flight Safety Foundation, 3–4 March 1989, Taipei, Taiwan.

Dinges, D.F., Mallis, M.M., Maislin, G., and Powell IV, J.W. (1998). *Evaluation of techniques for ocular measurement as an index for fatigue and as a basis for alertness management.* Washington, DC: US Department of Transportation.

Dinges, D.F., and Powell, J.W. (1985). Microcomputer analyses of performance on a portable, simple visual RT task during sustained operations. *Behavioural Research Methods, Instruments and Computers, 17*, 652–55.

Dinges, D.F., and Powell, J.W. (1988). Sleepiness is more than lapsing. *Sleep Research, 17*, 84.

Dorrian, J., Baulk, S.D., and Dawson, D. (2011). Work hours, workload, sleep and fatigue in Australian Rail Industry employees. *Applied Ergonomics, 42*(2), 202–9.

Dorrian, J., Hussey, F., and Dawson, D. (2007). Train driving efficiency and safety: examining the cost of fatigue. *Journal of Sleep Research, 16*(1), 1–11.

Dorrian, J., Lamond, N., Kozuchowski, K., and Dawson, D. (2008). The driver vigilance telemetric control system (DVTCS): investigating sensitivity to experimentally induced sleep loss and fatigue. *Behavior Research Methods, 40*(4), 1016–25.

Dorrian, J., Roach, G., Fletcher, A., and Dawson, D. (2007). Simulated train driving: Fatigue, self-awareness and cognitive disengagement. *Applied Ergonomics, 32*(2), 155–66.

Dorrian, J., Roach, G.D., Fletcher, A., and Dawson, D. (2006). The effects of fatigue on train handling during speed restrictions. *Transportation Research Part F: Traffic Psychology and Behavior, 9*(4), 243–57.

Dorrian, J., Rogers, N.L., and Dinges, D.F. (2005). Behavioural alertness as assessed by psychomotor vigilance performance. In C. Kushida (Ed.), *Sleep Deprivation: Clinical Issues, Pharmacology, and Sleep Loss Effects* (pp. 39–70). New York: Marcel Dekker.

Ferguson, S.A., Paech, G.M., Dorrian, J., Roach, G.D., and Jay, S.M. (2011). Performance on a simple reaction time task: Is sleep or work more important for miners? *Applied Ergonomics, 42*(2), 210–13.

Ferguson, S.A., Thomas, M.J.W., Dorrian, J., Jay, S.M., Wessenfield, A., and Dawson, D. (2010). Work hours and sleep/wake behaviour in Australian hospital doctors: It's not only about work. *Chronobiology International, 27*(5), 997–1012.

Feyer, A., Williamson, A., and Rassack, N. (1992). *The Information Processing and Performance Test Battery*. Sydney: National Institute of Occupational Health and Safety.

Foret, J., and Latin, G. (1972). The sleep of train drivers: An example of the effects of irregular work hours on sleep. In W.P. Colquhoun (ed.), *Aspects of Human Efficiency*. London: English Universities Press.

Glenville, M., Broughton, R., Wing, A.M., and Wilkinson, R.T. (1978). Effects of sleep deprivation on short duration performance measures compared to the Wilkinson auditory vigilance task. *Sleep, 1*(2), 169–76.

Grant, J.S. (1971). Concepts of fatigue and vigilance in relation to railway operation. *Ergonomics, 14*(1), 111–18.

Härmä, M., Sallinen, M., Ranta, R., Mutanen, P., and Muller, K. (2002). The effect of an irregular shift system on sleepiness at work in train drivers and railway traffic controllers. *Journal of Sleep Research, 11*(2), 141–51.

Harrison, Y., and Horne, J.A. (1997). Sleep deprivation affects speech. *Sleep, 20*(10), 871–7.

Hart, S.G., and Staveland, L.E. (1988). Development of NASA-TLX (Task Load Index): results of empirical and theoretical research. In P.A. Hancock and N. Meshkati (eds.), *Human Mental Workload* (pp. 239 50). Amsterdam: North Holland Press.

Hartley, L.R., Horberry, T., Mabbott, N., and Krueger, G.P. (2000). *Review of fatigue detection and prediction technologies.* Melbourne, Victoria: National Road Transport Commission. Available at http://www.ntc.gov.au/filemedia/Reports/ReviewFatigueDetectionandPredict.pdf.

Helmreich, R.L., Klinect, J.R., and Wilhelm, J.A. (2001). System safety and threat and error management: The line operational safety audit (LOSA). In R.S. Jensen (ed.), *Proceedings of the Eleventh International Symposium on Aviation Psychology* (pp. 1–6). Columbus, OH: Ohio State University.

Hockey, G.R.J., Healey, A., Crawshaw, C.M., Wastell, D.G., and Sauer, J. (2003). Cognitive demands of collision avoidance in simulated ship control. [Artcile]. *Human Factors, 45*(2), 252–65.

Horberry, T., Hartley, L., Krueger, G.P., and Mabbott, N. (2001). *Fatigue detection technologies for drivers: a review of existing operator-centred systems.* Paper presented at the People in Control: An International Conference on Human interfaces in Control Rooms, Cockpits and Command Centres, 19–21 June 2001, Manchester, UK.

Howlett, and Pudney, P. (2000). *Energy-efficient driving strategies for long-haul trains.* Paper presented at the CORE 2000 Conference on Railway Engineering, 21–23 May 2000, Adelaide, South Australia.

Hughes, J.H., and Parkes, S. (2003). Trends in the use of verbal protocol analysis in software engineering research. *Behaviour & Information Technology, 22*(2), 127–40.

Itoh, K., Arimoto, M., and Akachi, Y. (2002). *Gaze Relevance Metrics for Safe and Effective Operations of High-Speed Train: Application to Analysing Train Drivers' Learning with New Train Interface.* Tokyo: Tokyo Institute of Technology.

Johnson, L.C. (1982). Sleep deprivation and performance. In W.B. Webb (ed.), *Biological Rhythms, Sleep and Performance* (p. 111). New York: John Wiley & Sons, Inc.

Kleitman, N. (1963). Deprivation of sleep. In N. Kleitman (ed.), *Sleep and Wakefulness* (pp. 215–29). Chicago: University of Chicago Press.

Kogi, K., and Ohta, T. (1975). Incidence of near accidental drowsing in locomotive driving during a period of rotation. *Journal of Human Ergology (Tokyo), 4*(1), 65–76.

Lal, S.K.L., and Craig, A. (2001). Electroencephalography activity associated with driver fatigue: Implications for a fatigue countermeasure device. *Journal of Psychophysiology, 15*(3), 183–9.

Lamond, N., and Dawson, D. (1999). Quantifying the performance impairment associated with fatigue. *Journal of Sleep Research, 8*, 255–62.

Lamond, N., Dorrian J., Kozuchowski, K., Hussey, F.J., and Dawson, D. (2005). *Neurocom Driver/Operator Alertness Control System and Eyecheck Infrared Pupillometer Device: Validation Study.* Adelaide, Australia: Centre for Sleep Research, University of South Australia.

Lauber, J.K., and Kayten, P.J. (1988). Sleepiness, circadian dysrythmia, and fatigue in transportation accidents. *Sleep, 11*(6), 503–12.

Lieberman, H.R., Tharion, W.J., Shukitt-Hale, B., Speckman, K.L., and Tulley, R. (2002). Effects of caffeine, sleep loss, and stress on cgnitive performance and mood during US Navy SEAL training. Sea-Air-Land. *Psychopharmacology (Berl) 164*(3), 250–61.

Lieberman, P., Protopapas, A., and Kanki, B.G. (1995). Speech production and cognitive deficits on Mt. Everest. *Aviation, Space and Environmental Medicine, 66*(9), 857–64.

Loh, S., Lamond, N., Dorrian, J., Roach, G.D., and Dawson, D. (2004). The validity of psychomotor vigilance tasks of less than 10 minutes duration. *Behavior Research Methods, Instruments & Computers, 36*(2), 339–46.

Magill, R.A., Waters, W.F., Bray, G.A., Volaufova, J., Smith, S.R., Lieberman, H.R., … Ryan, D. H. (2003). Effects of tyrosine, phentermine, caffeine Damphetamine, and placebo on cognitive and motor performance deficits during sleep deprivation. *Nutritional Neuroscience, 6*(4), 237–46.

McNair, D.M., Lorr, M., and Droppleman, L.F. (1971). *Manual for the profile of mood states*. San Diego, CA: Educational and Industrial Testing Service.

Menzies, H. (2005). *No Time: Stress and the Crisis of Modern Life*. Vancouver/Toronto/Berkeley: Douglas & McIntyre.

Moller, H.J., and von Zerssen, D. (1995). Self-rating procedures in the evaluation of antidepressants. *Psychopathology, 28*, 291–306.

Naweed, A., Hockey, G.R.J., and Clarke, S.D. (2009). Enhanced information design for high-speed train displays: Determining goal set operation under a supervisory automatcd braking system. In D. de Waard, J. Godthelp, F.L. Kooi and K.A. Brookhuis (eds.), *Human Factors, Security and Safety* (pp. 189–202). Maastricht: Shaker Press.

Oman, C.M., and Liu, A.M. (2007). *Locomotive In-Cab Alerter Technology Assessment*. Cambridge, MA: Massachusetts Institute of Technology.

Patrick, G.T., and Gilbert, J.A. (1896). On the effects of loss of sleep. *Psychological Review, 3*, 469–83.

Pearce, K. (1999). *Australian Railway Disasters*. Sydney, New South Wales, Australia: IPL Books.

Pilcher, J.J., and Coplen, M.K. (2000). Work/rest cycles in railroad operations: effects of shorter than 24-h shift work schedules and on-call schedules on sleep. *Ergonomics, 43*(5), 573–88.

Pollard, J. (1991). *Issues in Locomotive Crew Management and Scheduling*. Washington, DC: Federal Railroad Administration.

Pudney, P., and Howlett, P. (1994). Optimal driving strategies for a train journey with speed limits. *Journal of the Australian Mathematical Society, Series B, 36*, 38–49.

Rajaratnam, S.M.W., and Arendt, J. (2001). Health in a 24-h society. *The Lancet, 358*(9286), 999–1005.

Ratcliff, R., and Van Dongen, H.P.A. (2011). Diffusion model for one-choice reaction time tasks and the cognitive effects of sleep deprivation. *Proceedings of the National Academy of Sciences of the United States of America, 108*(27), 11285–90.

Reason, J.T. (1990). *Human Error*. Cambridge: Cambridge University Press.

Roach, G.D., Dawson, D., and Lamond, N. (2006). Can a shorter psychomotor vigilance task be used as a reasonable substitute for the ten-minute psychomotor vigilance task? *Chronobiology International, 23*(6), 1379–87.

Roach, G.D., Dorrian, J., Fletcher, A., and Dawson, D. (2001). Comparing the effects of fatigue and alcohol consumption on locomotive engineers' performance in a rail simulator. *Journal of Human Ergology (Tokyo), 30*(1–2), 125–30.

Roth, E. (2000). In S. Reinach and T. Raslear (eds.), *An Examination of Amtrak's Acela High Speed Rail Simulator for FRA Research Purposes*. Washington, DC: Federal Railroad Administration.

Samn, S.W., and Perelli, L.P. (1982). *Estimating Aircrew Fatigue: A Technique with Implications to Airlift Operations*. Wright-Patterson Air Force Base, OH: USAF School of Aerospace Medicine, Aerospace Medical Division.

Smith, P., Shah, M., and da Vitoria Lobo, N. (2000). *Monitoring Head/Eye Motion for Driver Alertness with One Camera.* Paper presented at the 15th International Conference on Pattern Recognition, 3–7 September 2000, Barcelona, Spain.

Sodhi, M., Reimer, B., and Llamazares, I. (2002). Glance analysis of driver eye movements to evaluate distraction. *Behavior Research Methods, 34*(4), 529–38.

Stephan, K., Hosking, S., Regan, M., Verdoorn, A., Young, K., and Haworth, N.L. (2006). *The relationship between driving performance and the Johns Drowsiness Scale as measured by the Optalert system*. Melbourne: Monash University Accident Research Centre.

Thomas, G.R., and Raslear, T.G. (1997). *The Effects of Work Schedule on Train Handling Performance and Sleep of Locomotive Engineers: A Simulator Study.* Washington, DC: Federal Railroad Administration.

Thompson, K., Rainbird, S., and Dawson, D. (2010). The nature of the beast?: Metropolitan train drivers' experience, perception and recognition of fatigue. In C. Sargent, D. Darwent and G.D. Roach (eds.), *Living in a 24/7 world: The impact of circadian disruption on sleep, work and health* (pp. 1–5). Adelaide, South Australia: Australasian Chronobiology Society.

Tichon, J., Wallis, G., and Mildred, T. (2006). *Virtual Training Environments to Improve Train Driver's Crisis Decision Making*. Visual Computational and Learning Group. Paper presented at SimTect 2006, 29 May–1 June 2005, Melbourne, Victoria, Australia.

Torsvall, L., and Åkerstedt, T. (1987). Sleepiness on the job: continuously measured EEG changes in train drivers. *Electroencephalography and Clinical Neurophysiology, 66*(6), 502–11.

Torsvall, L., and Åkerstedt, T. (1987). Sleepiness on the job: continuously measured EEG changes in train drivers. *Electroencephalography and Clinical Neurophysiology, 66*, 502–11.

Verwey, W.B., and Zaidel, D.M. (2000). Predicting drowsiness accidents from personal attributes, eye blinks and ongoing driving behaviour. *Personality and Individual Differences, 28*(1), 123–42.

Whitlock, A. (2002). *Driver Vigilance Devices: Systems Review.* Surrey, UK. Retrieved from http://www.opsweb.co.uk/tools/common-factors/PAGES/Record.aspx?id=1249.

Williamson, A.M., Feyer, A.M., Mattick, R.P., Friswella, R., and Finlay-Brown, S. (2001). Developing measures of fatigue using an alcohol comparison to validate the effects of fatigue on performance. *Accident Analysis and Prevention, 33*(3), 313–26.

Wright, N., and McGown, A. (2001). Vigilance on the civil flight deck: incidence of sleepiness and sleep during long-haul flights and associated changes in physiological parameters. *Ergonomics 44*(1), 82–106.

Wright, N.A., Stone, B.M., Horberry, T.J., and Reed, N. (2007). *A review of in-vehicle sleepiness detection devices.* Berkshire: UK Transport Research Laboratory.

Yuceturk, A.V., Yilmaz, H., Eqrilmez, M., and Karaca, S. et al. (2002). Voice analysis and videolaryngostroboscopy in patients with Parkinson's disease. *European Archives of Oto-Rhino-Laryngology, 259*(6), 290–93.

Chapter 10

Applying the Theories and Measures of Situation Awareness to the Rail Industry

Janette Rose
University of South Australia, Adelaide, Australia

Chris Bearman
Central Queensland University, Appleton Institute, Adelaide, Australia

Anne Maddock
Railcorp NSW, Sydney, Australia

Introduction

This chapter provides an introduction to some of the basic issues around situation awareness, some of the key theories and some of the ways that situation awareness can be measured in the context of rail operations. The first part of this chapter deals with theories of individual and shared situation awareness and will be of particular interest to the human factors practitioner who wishes to know more details about the different ways that people have thought about situation awareness. The first two sections of this first part will be useful for the reader who is interested in learning more about situation awareness but has a limited knowledge of human factors. The second part of the chapter deals with issues around measuring situation awareness and will be of interest to both human factors practitioners and those who are required to exercise due diligence in the implementation of a new technology. This second part of the chapter also introduces a new method of assessing subjective situation awareness (LETSSA) that is currently being developed by two of the authors.

PART 1 – THEORIES OF SITUATION AWARENESS

What is Situation Awareness

Situation awareness is a key concept that underpins the complex actions performed by train drivers and network controllers. In essence, situation awareness describes the way that people build and maintain an ongoing understanding of the environment in which they are operating. Decisions about the control of the train or train control system are underpinned by that operator's situation awareness.

Where an operator has incomplete or inaccurate situation awareness, serious errors can occur. One example of error resulting from degraded situation awareness is the Tilt Train accident in Queensland, Australia:

> At 2355 Eastern Standard Time on 15 November 2004, the diesel tilt train, *City of Townsville*, VCQ5, derailed on the Bundaberg to Gladstone line. VCQ5 derailed on the first of a series of 60 km/h curve speed restrictions while travelling at a recorded speed of 112 km/h. One of the major factors identified in the subsequent investigation report was that it was possible that the driver became disorientated and/or distracted from his driving task leading into the curve and subsequently failed to recognise his geographic proximity along the track (Australian Transport Safety Bureau, 2005). The investigation report goes on to identify a number of other contributing factors, including the absence of the co-driver, who was not in the driving cab at the time of the accident, and the possibility that the driver mistook an alarm for a 'mid-section' magnet to be that of a station protection magnet, further along the track.

Situation awareness also underpins people's ability to predict what will happen in the environment in which they are operating (Endsley, 1995b). For example, a train driver approaching a temporary speed restriction may perceive the relevant elements in the environment and understand them but must also be able to predict how much braking will be required in order to slow the train sufficiently, allowing for the terrain, track conditions, length and weight of the train and weather conditions.

Why is Situation Awareness Important

As more complex and increasingly sophisticated information and control systems are introduced into the train cab and train control room, the major challenge for the operator of these systems is how to maintain an understanding of the current state of the task, the way that the technology is assisting to control that task, and what will be happening in the immediate future. This challenge is, in essence, one of maintaining appropriate situation awareness in a multi-modal dynamic situation.

Situation awareness has been identified as a key concept in safety and has been found to have a considerable impact on task performance in a wide variety of industries. The link between situation awareness and task performance has been supported in motor vehicle driving (Gugerty, 1997), flying aircraft (Orasanu, 1995) and air traffic control (Mogford, 1997). Studies that have investigated the effects of a number of different cognitive processes (such as intelligence, working memory and spatial memory) on task performance have found that situation awareness has an influence over and above these other processes (Durso, Bleckley and Dattel, 2006). Situation awareness is also a vital element of train driver route knowledge, where drivers need to build and maintain an understanding of where they are on the track. Models of train driver route knowledge generally incorporate a model of situation awareness (e.g. Luther, Livingston, Gipson and Grimes, 2005).

Technology and Situation Awareness

The way that each technology in the operational environment is designed and the way it interacts with other technologies will determine how easy it is for the operator to:

- perceive the current situation using that technology
- understand the functioning of the particular technology itself
- predict what will occur in the near future Endsley (1995b).

The quantity and quality of information provided by a system and how that information is displayed to the operator can determine how easily the operator can maintain situation awareness (Endsley, 1995b). Therefore, interfaces must be designed with a good understanding of the principles of situation awareness (Endsley, 1995b).

The complexity of the technology and its interaction with other technologies will also impact upon situation awareness, as operators need to keep track of different components requiring attention and the interaction with other components. As the rate of change of information in these components increases, the difficulty of maintaining adequate situation awareness also increases (Endsley, 1995b).

Automation has a complex relationship with situation awareness. On one hand, providing automation can lead to a reduction in the need to keep track of multiple interacting components of the system. On the other hand, automation can lead to a misunderstanding of what the technology is doing and result in the operator having a reduced level of situation awareness and being out of the loop. This is exacerbated when the operator does not have a good understanding of the system (Endsley, 1995b; see also Chapter 2 of this book).

There are other factors in the operational environment which may be indirectly influenced by a new technology and thereby have an impact on situation awareness. For example, physical stressors (such as noise and fatigue) and/or psychological stressors (such as fear and uncertainty) can result in reduced situation awareness (Endsley, 1995b). A certain amount of stress can have a beneficial effect by increasing attention to the important elements of a task, but high levels of stress can have a detrimental effect by narrowing the field of attention and leading to premature decision-making (Endsley, 1995b).

When evaluating new technology then, it is important to determine the impact that the technology has on the operator's situation awareness, particularly if the new technology is designed to present information to the operator.

Situation Awareness Theories

The following section takes a closer look at what situation awareness is and how it relates to other cognitive processes.

Situation awareness is a somewhat broad concept and it has been conceptualised in different ways by different researchers depending on their particular approach. The most common approach is based on the information processing metaphor of cognitive psychology (e.g. Endsley, 1988), but situation awareness has also been considered from the viewpoint of activity theory (e.g. Bedny and Meister, 1999), and from an ecological perspective (e.g. Smith and Hancock, 1995b).

There has also been some debate about what the term situation awareness actually refers to. Some researchers argue that situation awareness is the *process* of becoming aware (e.g. Fracker, 1991), while others argue that situation awareness is the *product* of situation assessment (e.g. Endsley, 1995b). Still other researchers argue that situation awareness is both a process and a product (e.g. Smith and Hancock, 1995b). When discussing the concept of situation awareness then, it is important to recognise that situation awareness may refer to the process of becoming aware, the product of that process or both.

To understand the concept of situation awareness in detail, it is necessary to consider the different ways that different researchers have thought about situation awareness and how they have represented its relationship to other similar processes. The following sections consider the different theories of situation awareness. The tilt train accident, described in 'What Is Situation Awareness' above, will be used as an example in the discussion of each of the situation awareness theories.

Endsley's Model of Perception, Comprehension and Projection

The most widely used model of situation awareness is that proposed by Endsley (1988) who defined situation awareness as 'the perception of the elements in the environment within a volume of time and space, the comprehension of their meaning, and the projection of their status in the near future' (p. 792). This definition is broken down into three levels: perception, comprehension and projection of future states.

According to Endsley's (1988) model, an operator must first perceive the various elements within the environment relative to their task, then bring together that information in short-term working memory to form a mental model. Then they must process that information to understand what it means in terms of their current goal. This then leads to a projection of what will happen in the future depending on the action they choose to take. It is not only the perceived elements

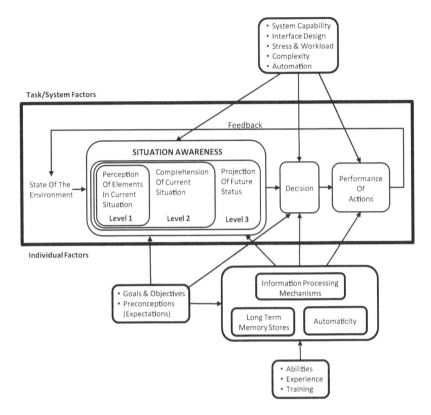

Figure 10.1 **Model of situation awareness in dynamic decision-making (Endsley, 1995b, reprinted by permission of Sage Publications, Inc.)**

in the environment that are used in working memory. An operator will draw information from their long-term memory (e.g. route knowledge) to assist in their level 2 (comprehension) and level 3 (projection) situation awareness and this will ultimately determine their performance. This model is illustrated in Figure 10.1, in the context of dynamic decision-making.

In Endsley's model, situation awareness is a knowledge state rather than a process (Endsley, 1995b). Endsley (1995b) refers to the process of achieving situation awareness as 'situation assessment'. In Endsley's (1995b) model, there are several factors that influence situation awareness. For instance, individual factors such as an operator's abilities, experience and training, may affect a person's information processing mechanisms. This, in turn, may affect goals, objectives and expectations which, in turn, influence situation awareness. In complex environments, such as control rooms, there may be many elements requiring an operator's attention thus their ability to manage this overload of information will impact upon their situation awareness (Endsley, 1995b).

A Practical Example of Endsley's (1988) Model of Situation Awareness

In the common situation where a train driver is approaching a yellow signal, the driver must:

- see the yellow signal (perception)
- understand that this means the next signal may be red (comprehension)
- think about the actions required to stop the train before the next signal (projection).

In this example, the driver's goal is to slow the train in preparation for stopping at the next signal, in case it is red.

The driver will actively seek out any information required to meet their goal and will interpret this information with their goal in mind (Endsley, 1995b). In this instance, the driver must actively seek information about:

- the current speed of the train
- the terrain leading up to the signal that might be red (e.g. track gradient).

The driver will use this information to form an understanding of what type and level of braking to use, i.e. automatic and/or dynamic (comprehension). They must then make judgements on whether the train is going to stop in time (projection of future states).

As the train progresses, the driver will continue to update their situation awareness by monitoring speed, deceleration rate and gauges, as well as using tactile senses such as 'feeling' the train slowing down (all considered to be perception). The driver will bring this information together to form a mental picture (comprehension) and will then think about whether the train is slowing down fast enough to be able to stop in time if the signal is red (projection of future states).

As can be seen by this example, situation awareness is not static but is constantly changing and evolving (Endsley, 1988).

An individual's situation awareness at all levels is relative to their goal or goals (Endsley, 1995b). For example, a freight train driver aiming to convey goods to the final destination in a safe and efficient manner will need to be aware of elements directly relevant to that goal, such as track gradient, speed limits and signal aspects. The driver will actively seek out any information required to meet their goal and will interpret this information with their goal in mind (Endsley, 1995b).

Experience and training can improve situation awareness by developing expectations about information that leads to faster and more accurate perception of the relevant information (Endsley, 1995b). Information gathered from the environment must be stored and analysed within working memory at the same time

---

The Tilt Train Accident according to Endsley's (1988) Model of Situation Awareness

According to Endsley's model of situation awareness, a number of factors may have led to the driver having degraded situation awareness in the lead-up to the accident.

Perception

- The environmental conditions (darkness) hindered the ability of the driver to perceive all relevant information, i.e. the speed warning board and other environmental cues that may have provided the driver with increased awareness of his exact location.
- There were no advanced speed warning boards, therefore the driver did not have an additional cue to forewarn him of the upcoming reduction in speed.
- The co-driver was not located in the cab immediately before the accident and so was not able to provide any additional cues to the driver.

Comprehension

The driver's comprehension of the situation was flawed due to his incorrect/incomplete perception of the cues in his environment. In addition, it is possible that the driver mistook the alarm tone of the 'mid-section' magnet to be that of the station protection magnet.

Projection

The driver had good route knowledge of the track layout and geometry, however, he incorrectly applied this knowledge to the projection of the future state, because the future state projection was based on an incorrect comprehension of his current state/location.

To summarise, the driver did not slow the train down because he did not perceive (and therefore comprehend) a number of important cues from the operating environment (speed sign/external environment). This resulted in incorrect or incomplete comprehension which then led directly to an incorrect projection of the future state of the environment, and therefore an incorrect action from the driver.

---

as projection of future states and decision-making, thus working memory plays a significant role in Endsley's concept of situation awareness (Endsley, 1995b).

Endsley (1995b) suggests that working memory may become congested for novices or experienced operators dealing with a novel situation. Experience and training help individuals to develop schemas and mental models for different situations which are stored in long-term memory and can be retrieved when required to assist in understanding the current situation and projecting future states (Endsley, 1995b). Experience can also lead to automaticity whereby certain actions are carried out without conscious thought and thereby reduce the demand on working memory (Durso and Dattel, 2006).

Endsley's (1988) model is intuitive and simple. It can be applied to any task in any domain and has, in fact, been used in a broad range of domains. With the three hierarchical levels of situation awareness, it is possible to easily and effectively measure situation awareness, and this model can be used to aid design and training aimed at the specific levels of awareness (Salmon et al., 2008). Another positive of this model is that it incorporates the numerous elements that influence the acquisition and development of situation awareness (Salmon et al., 2008). The idea of operators having mental models which develop with experience and lead to higher levels of situation awareness effectively explains differences between novices and experts (Salmon et al., 2008). This model has also been extended to explain team or shared situation awareness (see 'Shared Situation Awareness' later in this chapter).

Endsley's (1988) model is not without its critics. Smith and Hancock (1995b) argue that the incorporation of mental models is problematic as mental models are an ill-defined phenomenon. It also incorporates other psychological constructs, such as attention and schemata, that are not yet fully understood (Uhlarik and Comerford, 2002). Uhlarik and Comerford (2002) argue that the model fails to take into account the dynamic nature of situation awareness and Salmon et al. (2008) have questioned whether situation awareness is a separate construct to working memory in Endsley's model, given the similarity of the two concepts. Salmon et al. (2008) have also argued that the model is not entirely consistent with the stance that situation awareness and situation assessment are two separate constructs because the three levels appear to be processes rather than states, i.e. '*perception* of the elements', '*understanding* of their meaning' and '*projection* of future states'. However, despite these criticisms, Endsley's model remains the dominant model of situation awareness.

Activity Theory

Bedny and Meister (1999) use activity theory to define situation awareness. Activity theory has three basic structural components:

• Goals which represent the ideal image or logical result of activity
• Motives that direct individuals towards their goals
• Methods of activity that permit individuals to achieve their conscious goals.

According to this model, individuals are motivated by the difference between the current situation and their goal and the greater this difference, the more motivated they will be (Bedny and Meister, 1999). This process can be consciously driven or can be automatic (Bedny and Meister, 1999). Motivation is also affected by the significance of the goal, the significance of the consequences of failing to achieve that goal and obstacles to the achievement of the goal (Bedny and Meister, 1999).

There are three components/stages of activity: orientational, executive and evaluative (Bedny and Meister, 1999):

- In the orientational stage, individuals form a mental model of the world which leads to anticipation of the future state of the situation.
- In the executive stage, individuals make decisions and perform actions which transform the situation and move them closer to their goal.
- In the evaluative stage, information feedback leads to assessment of the result which then leads to corrective action (where necessary) and which may, in turn, influence the first two components.

Bedny and Meister's (1999) functional model of *orientational* activity is shown in Figure 10.2 where each block plays a role in the development and maintenance of situation awareness, and all function blocks are interconnected. The benefits of this model are that it explains the dynamic nature of situation awareness and describes the two-way interaction between situation awareness and the world (Salmon et al., 2008). The explication of the roles played by the various functions in the process of situation awareness is also useful for explaining their influence (Salmon et al., 2008).

On the downside, this model has received little attention and has very little, if any, empirical support (Salmon et al., 2008). In addition, it has not been extended to explain team or shared situation awareness, and because it incorporates both product and process, it is difficult to measure situation

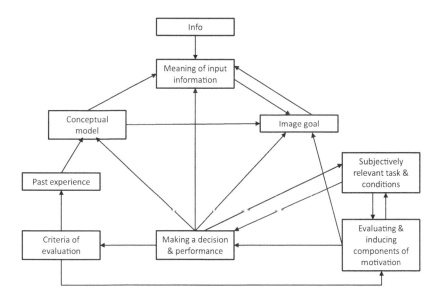

Figure 10.2   Functional model of orientational activity (adapted from Bedny and Meister, 1999)

The Tilt Train Accident According to Bedny and Meister's (1999) Activity Theory Model of Situation Awareness

In the context of driving the tilt train, the decision to reduce the speed was dependent on the ability of the driver to form an accurate mental model of where the train was on its journey.

The driver had an incorrect mental model (the orientational stage) of his location and therefore his *orientation* was incorrect. Due to the incorrect mental model the driver did not identify the need to reduce the speed of the train, i.e. he did not form a goal.

Without the identification of a goal there was no motivation for the driver to move from the current state (travelling at 112 km/h) to the goal state (travelling at 60 km/h). Therefore, the driver did not act to slow the train down on approach to the reduced speed limit, until the last minute, when the driver's orientation was belatedly correct.

Had there been additional cues to the driver, such as the presence of advanced speed restriction lineside signage or the advice of the co-driver, those cues may have supported the driver to form an accurate mental model, and therefore a goal to reduce the speed of the train.

---

awareness based on this model (Salmon et al., 2008). Salmon et al. (2008) argue that there is no link between 'image goal' and 'evaluative and inducing components of motivation', and that 'making a decision and performance' has no link to the world, which raises doubt about the ecological validity of this model. It is also a very complex model, making it difficult to understand and apply to specific tasks.

Perceptual Cycle Models

Adams, Tenney and Pew
Adams, Tenney and Pew (1995) argue that situation awareness can be seen as both product and process and both are interdependent and equally important. According to Adams et al. (1995), situation awareness as a product is derived from the process of gathering and interpreting information, and this process of situation awareness is guided by existing knowledge as well as expectations, hypotheses and familiarity with the situation.

Adams et al. (1995) use Neisser's (1976) perceptual cycle model to demonstrate the interdependent nature of the process and product of situation awareness (see Figure 10.3). As can be seen in this figure, an individual's mental model (schema) of the current situation guides their perceptual exploration of the environment which leads them to sample the available information, which may in turn impact upon their mental model (Adams et al., 1995). Put simply, the person's attention is directed to information believed necessary for the task

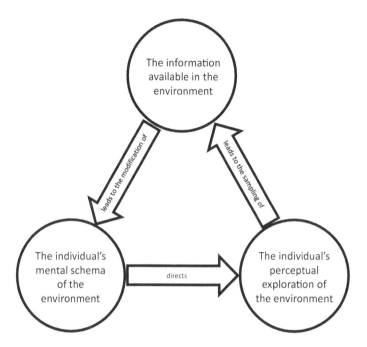

Figure 10.3   The perceptual cycle model (adapted from Neisser, 1976)

A Practical Example of Adams et al.'s (1995) Perceptual Cycle Model

Due to their route knowledge, a train driver will expect to see an upcoming signal and this will guide their perceptual exploration to search for that signal. At a cognitive level, there may be more than one possible interpretation of events and consideration of those potential interpretations leads to different expectations (Adams et al., 1995).

For example, when a driver is approaching a section of track where a signal is expected, the driver will have a schema of the present environment which includes the location of that signal. This will lead the driver to explore the environment to locate the signal to find out which of several possible events will be indicated (green to continue, yellow to slow or even potentially a malfunctioning signal which will require the train to be stopped and the controller to be contacted). If the signal is malfunctioning, the information available in the actual present environment will be lacking and will then affect the schema of the present situation. A driver may then look to their broader cognitive map of the world and its possibilities to find the reason for not being able to see the signal, which may include several possibilities (e.g. faulty memory of track, trees or other objects obscuring the signal), which could then lead to further action such as continued perceptual exploration or slowing of the train.

The Tilt Train Accident according to Adams et al.'s (1995) Perceptual Cycle Model

As stated above, the mental model (schema) of an environment and the expectation of future states based on that environment directs perceptual exploration. Given that the driver did not have an accurate mental model of his location, it is likely that he was no longer actively seeking information from his surroundings to confirm his understanding. Hence the situation awareness of the tilt train driver (as a product) immediately prior to the accident was incorrect.

It is understood from the accident investigaton that the driver had route knowledge of the area. Therefore, when the driver believed he was at a specific location along the track, it is likely that he would no longer (consciously) seek further information to confirm/clarify his situation awareness.

at hand, which leads to sampling of the information available in the current environment. When this information is incorporated into the mental model, the model is modified such that it then seeks different or further information and events (Adams et al., 1995).

The current state of the mental model is situation awareness as product, the state of the cycle at any particular moment in time is situation awareness as process, and the interaction between the two is product *and* process (Adams et al., 1995). All this falls within the inner circle of Neisser's perceptual cycle model. The outer circle of Neisser's model (see Adams et al., 1995, p. 89) represents a broader cycle in which the individual has access to their knowledge of the wider world, which may direct them towards actions aimed at gathering information that may not be present in the immediate environment.

Externally Directed Consciousness
Smith and Hancock (1995b) support Adams et al.'s (1995) definition and generally define situation awareness as 'adaptive, externally directed consciousness'. Smith and Hancock (1995a) argue that situation awareness is 'the capacity to direct consciousness in the generation of competent performance in particular situations' (p. 1). Put simply, operators choose what to pay attention to in order to perform specific tasks in specific situations. This is illustrated in Figure 10.4. This figure shows that to achieve situation awareness, a person's consciousness is either directed outwards, towards their environment, or inwards, to draw on their knowledge and experiences. The operator must aim to match goals, beliefs and knowledge with the specified task and performance criteria (Smith and Hancock, 1995b). When an operator's knowledge is not sufficient to match the state of the environment, they look for and acquire the data necessary to successfully achieve their goals. Thus situation awareness is 'the capacity to direct consciousness to generate competent performance given a particular situation as it unfolds' (Smith and Hancock, 1995b, p. 138).

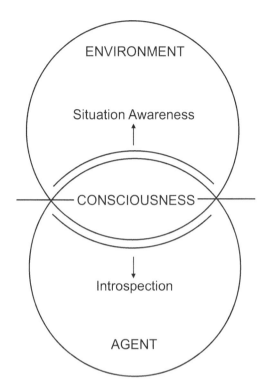

Figure 10.4 Smith and Hancock's (1995b) approach to defining situation awareness. (Reprinted by permission of Sage Publications, Inc.)

Smith and Hancock (1995b) argue that there are a number of constraints to the achievement of situation awareness. Specifically, an individual must have experience in the relevant environment and must have a knowledge store of suitable potential courses of action. They must also be able to assess performance variables in order to be able to modify their actions where performance is not optimal. Figure 10.5 illustrates these constraints.

Based on Smith and Hancock's (1995b) model, the arbiter of performance is found in the environment, rather than within the individual. Smith and Hancock (1995b) state that if goals and performance criteria were within the individual, they would always have perfect situation awareness because their perception would essentially be their goal. To achieve situation awareness, the operator must aim to match goals, beliefs and knowledge with the specified task and performance criteria (Smith and Hancock, 1995b). Although the arbiter of *performance* is in the environment, Smith and Hancock (1995b) argue that *situation awareness* is not solely within the person or the environment but it is the interaction between the two that is important. An operator's mental model of a situation leads them to

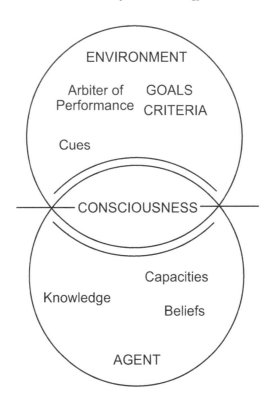

Figure 10.5   Constraints on situation awareness (Smith and Hancock, 1995b, reprinted by permission of Sage Publications, Inc.)

anticipate events and thus directs their attention to elements in the environment required to assist them in choosing a course of action (Smith and Hancock, 1995b). The situation is monitored to ensure that events progress as expected, with unexpected events leading the operator to search the environment further for an explanation, which will in turn modify their mental model and so on in a cyclical fashion (Smith and Hancock, 1995b).

As can be seen, this model suggests that situation awareness is a process of gathering information to arrive at a knowledge state, i.e. a process of situation awareness leading to a state of situation awareness.

According to Salmon et al. (2008), the two perceptual cycle models provide the most inclusive description of situation awareness, both in terms of achievement and maintenance. They are also inclusive of product and process and explain the dynamic nature of situation awareness (Salmon et al., 2008). However, both of these models make it difficult to measure situation awareness, either as a process or as a product (Salmon et al., 2008; Uhlarik and Comerford, 2002). As with the information processing approach, the perceptual cycle approach incorporates

---

The Tilt Train Accident According to the Externally Directed Consciousness Model (Smith and Hancock, 1995b)

Smith and Hancock's (1995b) model states that when an operator's knowledge is not sufficient to match the state of the environment, they will look for the data necessary to achieve their goals and will direct their attention to elements in the environment required to assist them in choosing a course of action.

In the tilt train incident, the driver had an inaccurate perception of where his train was located on the track and thus it is likely that the driver was no longer directing his attention to elements in the environment to confirm his understanding. Hence the driver did not take the appropriate course of action (i.e. he did not reduce the speed of the train) until immediately prior to the accident when further information, the speed sign, was available to him.

It is likely that the situation was not being monitored by the driver to ensure that events were progressing as expected because he did not have any goal-related expectations and therefore did not anticipate unexpected events which may have led him to search the environment further.

---

poorly understood psychological constructs, such as attention and schema, which is potentially problematic (Uhlarik and Comerford, 2002).

Taylor's Theory of Situation Awareness

Taylor (1990) describes situation awareness as having three major dimensions: *demands* on attentional resources; *supply* of attentional resources; and *understanding* of the situation. His theory was developed from data collected from military flight crews regarding tactical decision-making for pilots and navigators.

Ten generic concepts are embedded in the three dimensions, as follows:

Demands on attentional resources:

- Instability (the likeliness of sudden changes to the situation)
- Complexity (the level of complication or number of closely connected parts of the situation)
- Variability (the number of variables requiring the operator's attention).

Supply of attentional resources:

- Arousal (sensory excitability or the degree to which the operator is prepared for activity)
- Concentration (the degree to which the operator's thoughts are brought to bear on the situation)

The Tilt Train Accident according to Taylor's (1990) Model

In the context of Taylor's (1990) model of situation awareness, the factors that influenced the driver's actions in the tilt train accident can be described as follows:

Demands on attentional resources:

1. Instability: the situation would not have been considered to be unstable by the driver as there were limited variables.
2. Complexity: the situation was relatively straightforward. The driver was familiar with the route and train.
3. Variability: due to the operator's incorrect mental model, he did not expect to have to apply further resources to identify his location or act to manage the train.

Supply of attentional resources:

4. Arousal: it is possible that the driver may have been experiencing fatigue, given the time of the incident (2355) and was not therefore fully alert.
5. Concentration: based on investigation findings, it is likely that the driver's level of concentration on the driving task was reduced due to a number of factors – the driver was waiting for a drink to be made for him (from the co-driver); the driver may have been preparing to eat (sandwiches were found in the cab); the driver believed he was approaching a section of track that did not require active management of the train controls.
6. Division of attention: as above, a number of factors may have been dividing the driver's attention at the time of the incident.
7. Spare capacity: it is likely that the driver had spare capacity to address any changes to the environment that were presented to him, however, he did not have access to the information until immediately prior to the curve/speed reduction.

Understanding of the situation:

8. Information quantity: there were no advanced warning boards on approach to the curve at the time of the incident; the co-driver was not in the cab to provide confirmation of the driver's actions.
9. Information quality: the external cab environment was not visible to the driver as it was dark outside. In addition, there was no differentiation between the tone used for a 'mid-section' magnet alarm and the station protection magnet alarm.
10. Familiarity: the driver was familiar with both the route and the train, however, he had formed an incorrect mental model of his location.

Based on the parameters of this model, the combination of a reduced supply of attentional resources and a lack of prompts to support the driver's mental model of the situation led to the driver having reduced situation awareness.

- Division of attention (the degree to which the operator's attention is distributed or focussed)
- Spare capacity (the amount of cognitive ability available to apply to new variables).

Understanding of the situation:

- Information quantity (the amount of data received and understood)
- Information quality (the value of information received)
- Familiarity (the degree to which the situation is familiar) (Taylor, 1990).

There has been considerable criticism of Taylor's (1990) definition of situation awareness. It does not describe the processes involved in acquiring and maintaining situation awareness, and the measurement developed from Taylor's theory (SART: see description in the second part of this chapter) has not performed well in validation studies (Salmon et al., 2008). Further, Endsley (1993) argues that supply and demand of resources are related to workload and the only dimension directly related to situation awareness is understanding of the situation.

Hourizi and Johnson's Human-Computer Interaction Model

Hourizi and Johnson (2003) propose a predictive model of situation awareness developed with the aim of reducing breakdowns in human-computer awareness in situations where information is available but overlooked.

Although their model is based on the interaction between a pilot and their autopilot, the model is relevant to any task where there is communication between an operator and computer systems, e.g. between a signaller and their signalling control system. It could also feasibly be applied between an operator and any element of the environment or the environment as a whole (see example in the context of using TPWS below). Hourizi and Johnson (2003) argue that, before information can be seen as awareness, it must first pass through several cognitive processes. In this model, awareness arises when information from the environment is available (Level 1), has been perceived (Level 2), has been attended to (Level 3) and has undergone higher level cognitive processing (Level 4) (Hourizi and Johnson, 2003).

Hourizi and Johnson (2003) suggest that this model can be used in the early stages of the design process to make predictions about potential problems, as well as examining causes of breakdown in awareness.

Unlike Endsley's (1988) model where it is possible to have situation awareness at Level 1 (perception) only or at Level 2 (comprehension following perception), Hourizi and Johnson's (2003) model posits that all four levels must exist in order for an operator to have awareness (e.g. if an operator has Levels 1, 2 and 3 but not Level 4, they do not have situation awareness).

The Tilt Train Accident According to Hourizi and Johnson's (2003) Human-Computer Interaction Model

The control system in the tilt train presented the driver with a 'mid-section' magnet alarm on approach to the reduced speed restriction, which the driver acknowledged. The alarm activated (Level 1), was perceived by the driver (Level 2) and attended to (Level 3).

The tone for the 'mid-section' magnet alarm was the same as the tone used for the station protection magnet. The control system therefore did not provide the driver with any guidance on the location of the train. In fact, it is possible that the timing of the alarm reinforced the driver's incorrect assumption that he was approaching a station.

Therefore, in the context of Hourizi and Johnson's (2003) model, the awareness breakdown of the driver occurred at Level 4 – the higher level cognitive processing that would have ensured the driver understood the meaning of the alarm.

This model has the potential to be useful in determining the specific areas where awareness has failed and thus enable targeting of that area for redesign or training and thereby save time that might otherwise be wasted. However it does not fully explain the processes involved in acquiring and maintaining situation awareness (Salmon et al., 2008).

Summary

The models of situation awareness presented in this section represent a sample of the most common models that have been proposed over the last two to three decades. Each set of researchers have brought their own perspective to situation awareness and the way that situation awareness relates to other similar processes. The debate continues as to whether situation awareness is a unique construct separate to other processes or whether it is really just an ability or part of other constructs such as working memory (Salmon et al., 2008).

Whatever the case may be, situation awareness has been linked to task performance as well as errors and accidents and thus is worthy of careful consideration when considering human factors issues relating to dynamic tasks and the changes that occur when introducing new technology. At a practical level, the particular model that is used to frame an investigation into situation awareness is determined by the purpose and context of the investigation and the preference of the individual. Each model provides a detailed account of situation awareness and a slightly different way of looking at the concepts, which frames the questions that are asked and the way the answers are understood.

Each of the models considered so far focuses on how an individual builds and maintains situation awareness, however teams also need to build and maintain a collective situation awareness in order to function effectively. The next section of this chapter considers shared situation awareness.

Shared Situation Awareness

Most operations in the rail industry are considered to be single-person tasks. Only one person can actually drive the train, even if there are two drivers physically present in the cab, and train control is typically based around a single controller who makes decisions and issues instructions. However, to produce effective performance, network controllers need to have an awareness of what is happening in other parts of the network, which requires developing a shared situation awareness with other controllers. Developing shared situation awareness is also important in two-driver operations since the second driver plays an important role in monitoring for and mitigating errors made by the first driver. The network controller and driver together also need to possess a shared situation awareness so that appropriate decisions can be made about the train in non-standard situations.

Increasingly, tasks are being carried out within a complex social structure where the effects of a change in a task for one operator affects the way that other operators need to conduct their tasks. When a new technology is introduced, it is important to consider the impact that it will have on the situation awareness of the individual, the teams that the individual operates within and the wider networks in which that person conducts their task.

A More Detailed Look at Shared Situation Awareness

Team Situation Awareness Models – Endsley (1995b)
One way of conceptualising shared situation awareness is at the intersection between the different elements of individual situation awareness. Endsley (1995b) expresses this concept in the model shown in Figure 10.6 below.

In Endsley's (1995b) model, each team member has their own role and needs to maintain situation awareness according to their individual requirements. Shared situation awareness with other team members occurs where there is overlap in the individual situation awarenesses. Thus each individual team member will be building and maintaining their own Level 1, 2 and 3 situation awareness and sharing this with other members of the team as appropriate.

Although sharing information is a vital element of shared or team situation awareness, Bolstad and Endsley (2000) found that information shared must only be that which is necessary for building shared mental models. They tested the effects of two types of shared displays as well as the effects of workload on the development of team situation awareness. One display provided the team members with all the information available to the other team member (Full Shared Displays), while the other display provided only information critical for shared information requirements (Abstracted Shared Displays). Bolstad and Endsley (2000) found

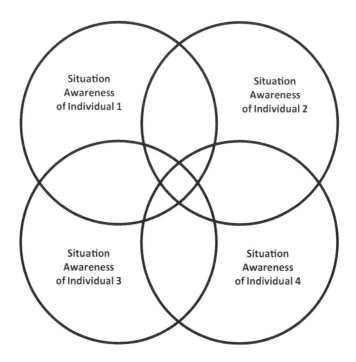

Figure 10.6   Team situation awareness (adapted from Endsley, 1995b)

Endsley's (1995b) Model of Team Situation Awareness in the Context of Passenger Trains

On US Amtrak passenger trains, the crew is made up of at least three individuals: the engineer (driver), the conductor and the assistant conductor (Roth and Multer, 2009). The engineer is in charge of running the train, including compliance with signals and speed restrictions; the conductor ensures that doors are closed and it is safe for the train to move; and the assistant conductor assists with fare and ticket collection and is responsible for passenger safety in emergency situations (Roth and Multer, 2009). Each person has their own individual situation awareness requirements but they also have overlapping requirements. The amount of overlap, and therefore the amount of team situation awareness, will vary depending upon the current tasks, goals and roles of the individuals in the team (Endsley, 1995b). For example, before departing a station, the conductor and the engineer must both be aware that all doors are closed and it is safe to move off.

Full Shared Displays vs Abstracted Shared Displays in the Train Control Centre

Train control centres often have an overview screen which provides visibility of the entire area of the network managed by that signalling complex. Signallers therefore have all the information associated with all the other signalling control panels presented to them simultaneously. In locations where an overview screen is not used, signallers are only provided with visibility of the adjacent areas of control, usually through the use of mimic screens located on their workstation.

In the context of Bolstad and Endsley's (2000) study, an overview screen in a signalling control centre could be considered to be a Full Shared Display and the mimic screens on the individual workstations could be considered to be Abstracted Shared Displays. This has interesting implications for the way in which the situation awareness of train controllers would best be supported.

While conclusions about the different ways of presentating information to train controllers should not be based on this finding alone, it does suggest that an assessment of shared situation awareness under different conditions of workload is an important part of the evaluation of changes to the design of control centres.

---

that the Full Shared Displays led to a reduction in overall performance, particularly under high workload conditions, as the full display added to overall workload. In contrast, the Abstracted Shared Displays resulted in improved performance, with the benefits increasing as workload increased. This benefit was due to the difficulty of direct communication between team members under high workloads, with the Abstracted Shared Displays providing only vital information and therefore not increasing workload (Bolstad and Endsley, 2000).

One of the strengths of Endsley's (1995b) theory of team situation awareness is that it is based on Endsley's (1988) theory of individual situation awareness which has been widely accepted, has a sound theoretical underpinning and is well-supported in the literature (Salmon et al., 2008). As with the individual situation awareness theory, this model can be applied in a wide variety of domains (Salmon et al., 2008) and is simple and easy to understand. It is possible to measure team situation awareness, based on this model, using SAGAT (Salmon et al., 2008). SAGAT is discussed further below in 'Situation Awareness Global Assessment Technique (SAGAT)'.

However, the model has been criticised as being a simplistic extension of Endsley's three-level model of individual situation awareness rather than a model of team situation awareness in its own right (Salmon et al., 2008). Although SAGAT can be used to measure team situation awareness based on this model, using SAGAT for this purpose is complex and has to be done using simulation (Salmon et al., 2008).

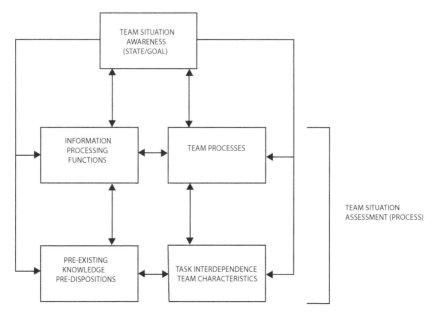

Figure 10.7   Conceptualisation of team situation awareness (Salas et al., 1995, reprinted by permission of Sage Publications, Inc.)

Team Situation Awareness Models – Salas, Prince, Baker and Shrestha (1995)
A more detailed model of shared situation awareness has been proposed by Salas, Prince, Baker and Shrestha (1995), as shown in Figure 10.7 above. The key principles of this model are that teamwork must involve two or more people working towards a common goal, each with a specific role to perform, and those people must work interdependently. Teams may also have shared mental models that assist in understanding and predicting other team members' behaviour (Salas et al., 1995).

According to Salas et al. (1995), the two vital elements of team situation awareness are individual situation awareness and team processes. Team processes include cognitive processes and behaviours that assist the performance of the team. Salas et al. (1995) refer to the process of achieving a state of situation awareness as 'team situation assessment'. Team situation assessment includes the pre-existing knowledge/predispositions of individuals and task interdependence/team characteristics. These different factors can interact with each other and either directly or indirectly influence team processes and team information processing (Salas et al., 1995).

Team situation awareness as a state or goal may directly and/or indirectly influence all the elements of the process (Sales et al., 1995). Sharing of information between team members can offset deficiencies that may be present in any individual's mental model in long-term memory (Salas et al., 1995). Communication between team members of factors relating to team performance,

---

Salas et al.'s (1995) Model of Team Situation Awareness in the Context of Relay Teams in Long-Haul Train Driving

In long-haul operations that cover great distances in Australia, train drivers work in relay teams whereby two drivers rest while the other two drivers drive the train and they change places at certain designated points along the route. Each driver has their own pre-existing knowledge and predispositions that affect the team characteristics. They all have the shared goal of arriving at their destination safely and in reasonable time. Both active drivers constantly monitor the environment, relevant gauges and the radio. They may share their perceptions of the external environment by cross-calling and the second driver will alert the first driver of any deficiencies in their mental model. For example, if the first driver does not begin to slow the train on an approach to a speed reduction due to a deficiency in their mental model, the second driver will share his knowledge and experience and thereby lead the driver to seek information from the environment to correct their mental model. At the time of changeover, the active drivers will update the new drivers on the current situation to ensure they are aware before taking control of the train.

---

such as goals and team capabilities, will affect perception of information from the environment (Salas et al., 1995). Good communication between team members should lead to a shared understanding of the current situation which, like individual situation awareness, is dynamic (Salas et al., 1995).

This model provides an understanding of the team processes relating to team situation awareness (Salmon et al., 2008). Rather than merely being an extension of individual situation awareness models, this model was based on the teamwork literature (Salmon et al., 2008). Salas et al. (1995) relate this model to the training of teams and propose what should be measured and how measurement can be conducted during assessments of team situation awareness (Salmon et al., 2008).

According to Salmon et al. (2008), however, the measurement approach proposed by Salas et al. (1995) is more appropriate for measuring team performance and behaviour rather than team situation awareness. The model focusses on the processes of teams rather than on their situation awareness (Salmon et al., 2008). The model also does not have a basis in naturalistic or empirical studies (Salmon et al., 2008).

Distributed Cognition Approach

A different way of looking at shared situation awareness is in terms of the whole system in which the work takes place, rather than it being just the product of individual cognitions (Artman and Garbis, 1998). The definition of situation awareness that occurs in teams is defined by Artman and Garbis (1998) as, 'The active construction of a model of a situation partly shared and partly distributed between two or more agents, from which one can anticipate important states in the near future' (p. 2).

Artman and Garbis (1998) argue that since most dynamic systems involve tasks carried out by teams, existing models are inadequate for the study of team situation awareness. They suggest that studying control in dynamic systems requires a shift in focus from individuals to the cognitive system as a whole, incorporating not only the people but also the artefacts they use. Their viewpoint is based on the theory of distributed cognition which focuses on the way that information is represented and how those representations are transformed and disseminated (Artman and Garbis, 1998). Artman and Garbis (1998) state that the overall cognitive system comprises the interactions that occur between individuals and between individuals and the system. These interactions include the resources and information provided by individuals (Artman and Garbis, 1998).

Operators in dynamic systems are often detached from the situation or environment they are aiming to control and thus have to rely on other operators and artefacts to provide the information necessary to successfully meet task goals (Artman and Garbis, 1998). Team members have different knowledge and different resources which need to be brought together in order to meet the overall goal of the team, and the sense-making process occurs throughout the team rather than in only one individual (Artman and Garbis, 1998).

This model describes situation awareness at a systems level and therefore allows assessment of individual, collaborative and systemic situation awareness (Salmon et al., 2008). This model also has sound theoretical underpinning (Salmon

---

Artman and Garbis's (1998) Whole System Model in the Context of the Rail Environment

A rail operator in New South Wales, Australia, manages train movements with the involvement of three parties:

- Train controller: responsible for decision-making regarding the train routing
- Signaller: responsible for setting the route for the train to travel through the network, conveying relevant information to the driver that may affect the way they manage their train
- Train driver: responsible for safely managing the train.

All parties have access to a number of systems to support their situation awareness, some of which overlap:

- Overview/mimic panel: train controller and network controller
- Lineside infrastructure: driver
- Signalling control system: network controller.

All parties use the information presented to them, in conjunction with communications (via the radio) to form their mental model of the situation and make decisions based on this information. The participants are therefore actively constructing a model of the situation that is 'partly shared and partly distributed between two or more agents' that is then used to 'anticipate important future states in the near future' (Artman and Garbis, 1998).

et al., 2008). According to Salmon et al. (2008), the weaknesses of this model are that its application is limited, there is no prescribed measurement approach and it does not describe the processes of individual situation awareness.

Distributed Situation Awareness Model

Stanton et al. (2006) propose a theory similar to that of Artman and Garbis (1998), which focuses on interactions between human and non-human agents in sub-systems. In this theory, the situation awareness of the various agents may be different to each other and the sharing of awareness is not always necessary or useful. Stanton et al. (2006) use Endsley's (1988) three-level model of situation awareness (perception, comprehension and projection) and apply it to a system involving a team.

This theory suggests that situation awareness is present at the system level rather than the individual level (Stanton et al., 2006). Six tenets are proposed to form the basis of this theory:

1. Human and non-human agents may possess situation awareness.
2. The views of the scene will vary between agents.
3. Overlapping of situation awareness between agents is dependent upon their respective goals.
4. Communication between agents may take several forms other than verbal, e.g. customs and practices, non-verbal behaviour.
5. Systems are held together by situation awareness.
6. Degraded situation awareness in one agent may be compensated for by another agent (Stanton et al., 2006).

Stanton et al. (2006) – Example of Shared Situation Awareness in Train Driving

When a driver is driving with a driver-tutor in a train with Automatic Train Protection (ATP):

- the ATP system monitors the speed of the train (perception), compares the current speed to the speed limit of the track (comprehension) and then provides an alert to the driver that they are overspeeding/not slowing down quickly enough
- the train driver receives the input from the ATP interface (perception), identifies the appropriate action to address the alert (comprehension) and applies the brake
- the driver-trainer sees the action of the train driver (perception), recognises that the train is about to slow down (comprehension) and braces himself against the deceleration (projection).

This example illustrates that knowledge is distributed across the system, and that information can be transferred implicitly rather than an explicit sharing of mental models (Stanton et al., 2006). Note that the first agent in this scenario (the ATP system) is an artefact not a person.

As with Artman and Garbis's (1998) approach, Stanton et al.'s (2006) model describes team situation awareness at a systems level, allowing assessment of individual, collaborative and system situation awareness (Salmon et al., 2008). It is also underpinned by sound theory and has been applied in numerous domains in which collaboration is required (Salmon et al., 2008).

According to Salmon et al. (2008), the main weaknesses of this model are that: the description and measurement of distributed situation awareness are subjective, often occuring post-task; the propositional networks methodology that is used for measuring situation awareness based on this model lacks validation; and the model does not describe the processes of individual situation awareness.

The theories of Artman and Garbis (1998) and Stanton et al. (2006) are referred to as distributed situation awareness rather than shared situation awareness. The main difference between shared situation awareness and distributed situation awareness is that shared suggests mutual requirements and purpose, whereas distributed suggests differing requirements and purposes that may be compatible (Stanton et al., 2006). In fact, compatability is essential for the successful performance of collaborative tasks (Salmon, Stanton, Walker, Jenkins and Rafferty, 2010).

Summary

As systems become more complex and as we increasingly recognise that work occurs in social contexts, the study of how shared situation awareness is affected by changes that occur due to the introduction of new technology becomes increasingly important. It is entirely possible that introducing a new technology may have no discernible effects on an individual's situation awareness but has large implications for the way that situation awareness is shared with others. Thus, in considering the implications of introducing a new technology, it is important to evaluate how the changes will influence both individual and shared situation awareness.

Part 1 Conclusions

We have considered the reasons why situation awareness is important, the different approaches that researchers have taken to understanding situation awareness and the issues surrounding shared situation awareness. It is clear that having good situation awareness makes good performance more likely and degraded situation awareness makes it more likely that errors will occur, although it should be noted that this is not a necessary relationship (i.e. errors will not always occur with degraded situation awareness). This means that it is important to carefully consider the implications for individual and shared situation awareness for any new technology. The second part of this chapter presents a review of a number of tools and techniques developed for measuring both individual and team/shared situation awareness and discusses the strengths and limitations of the different tools.

PART 2 – MEASURING SITUATION AWARENESS

The first part of this chapter outlined why situation awareness is important, the different ways that different researchers have thought about the concept of situation awareness, how it is related to other processes and the importance of considering shared situation awareness. In this part of the chapter, we turn to a consideration of how to measure situation awareness in rail operations.

There are two main reasons why an operations or human factors manager should be interested in measuring situation awareness:

1. A good situation awareness measure has the potential to be more sensitive than performance measures (Durso and Dattel, 2004). Situation awareness measures are likely to be more effective for detecting differences between operators, displays or systems.
2. It has been argued that situation awareness can predict potential performance problems even if no problems exist at the time (Durso and Dattel, 2004), that is, situation awareness can potentially detect 'accidents waiting to happen'.

There is a considerable number of different methods for measuring situation awareness. It is important to have a good grasp of the benefits and limitations of each method so that informed decisions can be made regarding the correct method or methods to use for the specific purpose required.

Key Terms Relating to Measurement Constructs

In order to examine these benefits and limitations we first need to introduce a few key terms. In human factors research there are certain expectations that relate to measuring constructs such as situation awareness. These are listed and described below (taken from Uhlarik and Comerford, 2002):

- *Validity*: the degree to which a measure actually measures that which it is intended to measure. There are different types of validity:
  - *Face validity*: On the surface, the measure appears to be measuring the psychological construct of interest.
  - *Construct validity*: The measure actually measures that which it intends to measure.
  - *Concurrent validity*: the level of correlation with other measures.
- *Sensitivity*: the degree to which a measure can differentiate between different conditions; for example, the measure would need to be able to detect differences in situation awareness that may be present when comparing a new information system with an old system.
- *Selectivity*: the degree to which the measure can detect changes to the

construct of interest without being affected by other constructs; for example, if the construct of interest is situation awareness, the investigator does not want the results confounded with other constructs such as workload.

- *Obtrusiveness*: the degree to which a measure influences the primary task; for example, the administering of a measure to assess a train driver's situation awareness should not impact upon their performance of that task, either positively or negatively.
- *Reliability*: the consistency of a measure; for example, if a train driver is tested in a simulator on more than one occasion using the exact same scenario and their situation awareness was exactly the same each time (which is highly unlikely but useful as an example), the results of the measure should be the same each time.

Having outlined some of the key ideas about evaluating research methods we now turn to a consideration of the methods themselves.

Measures of Situation Awareness

The next six sections provide a summary of the approach/method, validity and reliability, and pros and cons for each of the following situation awareness assessment tools:

- Situation Awareness Global Assessment Technique (SAGAT)
- Situation Present Assessment Method (SPAM)
- Situation Awareness Rating Technique (SART)
- Situation Awareness Subjective Workload Dominance (SA-SWORD)
- Situational Awareness Linked Indicators Adapted to Novel Tasks (SALIANT)
- Low-Event Task Subjective Situation Awareness (LETSSA).

Then the next four sections provide a summary of a number of supporting tools/ techniques that may also be used when measuring situation awareness, including:

- eye-tracking
- verbal protocol
- performance-based measures
- observer ratings.

Situation Awareness Global Assessment Technique (SAGAT)

One of the most common methods of assessing situation awareness is the Situation Awareness Global Assessment Technique (SAGAT). Endsley (1988) developed

SAGAT to provide a way of evaluating new concepts and technologies aimed at enhancing pilot situation awareness.

Approach
SAGAT involves freezing a simulation at randomly selected intervals and questioning operators about their perceptions of various elements of the situation at the time of the freeze. Questions relate to the three levels of Endsley's (1988) situation awareness model. Answers given by the participant are compared with the data provided by the simulator computer and a composite score is calculated.

Validity and Reliability
In a study conducted by Endsley and Bolstad (1994), two separate assessments of situation awareness using the same questions showed very similar results, which demonstrates that SAGAT has good test-retest reliability. A number of studies have shown that SAGAT can detect differences in situation awareness, suggesting it has good sensitivity. For example, Ma and Kaber (2007) found that as Level 3 situation awareness increased performance errors declined.

Some concern has been raised about the obtrusiveness of freezing simulations to administer questionnaires and the effect it may have on performance (Sarter and Woods, 1991, 1995). However, numerous studies have found that freezing does not affect performance (e.g. Endsley, 1995a; Endsley, 2000). Research by Endsley (1995a) and Gugerty (1997) has addressed concerns about the effect of memory on probe-recall measures of situation awareness such as SAGAT, and found that these measures are not hindered by retrospective recall or constraints of implicit memory, thereby further supporting the construct validity of SAGAT. SAGAT has also been found to have predictive validity in a study of fighter pilot performance in a combat simulation (Endsley, 1990).

As well as concern relating to the potential impact on performance, Sarter and Woods (1995) have argued that pausing the simulation and asking the operator questions is likely to impact upon their situation awareness. They argue that questions act as retrieval cues and so can affect the information that operators will recall to mind. SAGAT therefore is argued to measure the knowledge an operator can recall when prompted but not necessarily the knowledge that they would consider relevant themselves (Sarter and Woods, 1995). There is also an assumption that all knowledge used in a task is accessible from memory when the task is frozen when, in fact, it is possible to use knowledge without being aware of it (Selcon, Taylor and Koritsas, 1991).

Pros
- SAGAT incorporates all situation awareness requirements.
- SAGAT directly measures the operator's knowledge of the situation.
- SAGAT can be objectively collected and evaluated.
- SAGAT possesses direct face validity (Endsley, 1988) and has good construct validity (Fracker, 1991).

- SAGAT does not cause additional workload for the participant, unlike other methods (Endsley, 1995a).
- SAGAT allows collection of data regarding situation awareness at various points throughout the task (Endsley, 1988).

Cons

- Simulation must be used and the simulation must be paused to collect data (Endsley, 1988). This makes it unsuitable for use in uncontrolled environments (i.e. the 'real world').
- The act of prompting an operator to recall specific information is a retrieval cue and although it will shed light on the information an operator can recall when prompted, it will not necessarily show the knowledge a participant will reveal on their own or what they would see as relevant (Sarter and Woods, 1995).

Despite the limitations associated with SAGAT, it remains one of the most popular methods of assessing situation awareness and is one of the main comparison measures used to assess other measures.

Situation Present Assessment Method (SPAM)

Like SAGAT, SPAM uses probes to measure situation awareness, however, unlike SAGAT, SPAM does not require freezing of a simulation in order to present the queries nor does it require the blanking out or covering up of monitors (Durso and Dattel, 2004).

Approach

With SPAM, operators continue to perform their tasks while they answer the queries, which can relate to current and future situations (Durso et al., 1995). The accuracy of responses to SPAM queries is usually very high because the current situation is not concealed so the operator has access to the relevant information. The measure of situation awareness is therefore calculated on response time, measured from the time the operator actually looks at the query until they respond.

Reliability and Validity

SPAM was developed by Durso et al. (1995) who assessed five different methods of measuring situation awareness in comparisons between expert and novice chess players. Accuracy of responses to SPAM probes was not significantly different between experts and novices, as would be expected, however, the response times of experts were significantly faster than novices, suggesting a higher level of situation awareness (Durso et al., 1995), and therefore suggesting good construct validity.

In a study conducted by Durso et al. (1999), SPAM was one of three measures used to assess the situation awareness of air traffic control instructors. Two measures of performance were used: subject matter expert evaluations, and remaining action counts (the number of remaining actions required to complete the task). SPAM scores were predictive of performance on both measures, suggesting good reliability.

SPAM does not rely on the participant's memory since situation awareness is assessed on the basis of response time, which is argued to provide a more sensitive measure of situation awareness (Durso and Dattel, 2004). It is also argued that situation awareness is best when the operator is engaged and the situation is current (Durso and Dattel, 2004).

The timing of query presentation in SPAM could be problematic, as an operator's workload is likely to affect response times, i.e. if very busy, the operator will take longer to respond than if they have very little to do (Durso and Dattel, 2004). To overcome this problem, the operator is warned that a query is ready for answering and the query is not presented until the operator indicates their readiness to respond (Durso and Dattel, 2004).

Pros

- SPAM has good face validity and construct validity and allows situation awareness to be assessed when it succeeds, not just when it fails, since the measure is based on reaction times rather than memory (Durso and Dattel, 2004).
- Durso and Dattel (2004) argue that it is better to know where to look for information rather than use cognitive resources to remember it.
- Unlike SAGAT, SPAM can be used in the real world because it does not require information on the displays to be concealed (Durso and Dattel, 2004).

Cons

- SPAM has the potential to be influenced by workload (Durso and Dattel, 2004), however, careful construction of the research procedure should ensure that this does not occur and also that the measure does not create a distraction to the operator.

SPAM seems to be a useful addition to the situation awareness measures, with a potential application to real-world task performance.

Situation Awareness Rating Technique (SART)

The Situation Awareness Rating Technique (SART) is a subjective measure of situation awareness proposed by Taylor (1990) and developed using information from military flight crews.

Approach

Consistent with Taylor's (1990) theory of situation awareness (see 'Taylor's Theory of Subjective Situation Awareness' in the first half of this chapter), SART has 10 dimensions that can be grouped as follows:

- *Demands* on attentional resources (instability, complexity and variability of the situation)
- *Supply* of attentional resources (arousal, concentration, division of attention and spare capacity)
- *Understanding* of the situation (information quantity, information quality and familiarity) (Taylor, 1990).

SART is available in two forms (the 3-D SART and the 10-D SART) and is usually given as a questionnaire with responses being rated on a continuous or 7-point rating scale (Taylor, 1990). Where it is necessary to keep manual or visual interference to a minimum, the 3-D SART can be administered verbally such that participants respond to questions via verbal report (e.g. answering 'Low', 'Medium', or 'High') (Taylor, 1990). The 10-D SART takes longer to administer than the 3-D SART but provides a higher level of specificity and diagnostic power (Taylor, 1990).

In naturalistic environments, SART is usually administered during a convenient break and in simulation it is usually administered at the end of a scenario. Endsley, Selcon, Hardiman and Croft (1998) state that SART is suitable for any domain without the need for customisation, although Rose, Bearman and Dorrian (2012a, 2012b) report that SART does not apply very well to long-haul freight operations.

Reliability and Validity

Endsley et al. (1998) found that SART was highly correlated with subjective performance but not significantly correlated with SAGAT, a more objective measure of situation awareness. Testing conducted by Selcon and Taylor (1990) found that SART was correlated with performance measures, however Endsley (1995a) argues that it is unclear whether this correlation is due to workload or the 'understanding' components of SART. It is possible that different participants may have different interpretations of the concepts being measured, thus, two individuals having the same SART score does not mean that they actually had the same level of situation awareness (Fracker, 1991).

Endsley (1995a) has questioned whether SART measures situation awareness or whether it is in fact a measure of workload, which would suggest poor selectivity. An experiment conducted by Selcon et al. (1991) in which ratings on 3-D SART, 10-D SART and a workload measure were collected and comparisons made between experienced and inexperienced pilots found that 10-D SART was sensitive to experience but the workload measure was not, thus SART appears to measure something other than workload (Uhlarik and Comerford, 2002). However, Stanton, Salmon, Walker, Baber and Jenkins (2005) state that testing of

SART often reveals correlations between situation awareness and workload (e.g. Selcon et al., 1991). SART can also be strongly influenced by performance such that operators are likely to report good situation awareness when their performance is good and vice versa (Endsley, 1988; Endsley et al., 1998; Salmon, Stanton, Walker, and Green, 2006).

Despite these issues, Endsley et al. (1998) have suggested that measures such as SART are useful because there may be an important link between an operator's perceived situation awareness and their subsequent behaviour (i.e. how their perception of their situation awareness will influence their actions).

Pros
- SART can be used in naturalistic environments by both operators and observers. For example, SART can be completed by train drivers when they first change places with the second driver or when they are stopped at a station or other location.
- SART is non-intrusive and is quick and easy to administer (Salmon et al., 2006).
- SART has been argued to access participants' subjective ratings of situation awareness, which may have an important influence on performance not examined by other measures, such as SAGAT.

Cons
- SART may not be an accurate measure of situation awareness in that people may not admit to lapses in situation awareness (Hunn, 2001) or may not have insight into their situation awareness.
- The selectivity of SART has been questioned, since it appears to be influenced by both workload and performance.
- It is difficult to compare two different people's situation awareness using SART, since people may use different interpretations of the constructs to make their rating.

SART may be useful in providing a measure of subjective situation awareness, although given the problems with construct validity it should be used with caution.

Situation Awareness Subjective Workload Dominance (SA-SWORD)

SA-SWORD is a pair-wise comparison measure developed by Vidulich and Hughes (1991) and is a subjective rating measure used to compare subjective situation awareness between two different interfaces or systems.

Approach
Used mostly in the aviation domain, SA-SWORD requires operators to evaluate two different interfaces and give their subjective responses to questions regarding which of the two interfaces is the most effective at assisting them in various tasks.

Reliability and Validity

In a comparison of perceptual judgement and subjective measures of spatial awareness, Bolton and Bass (2009) found that SA-SWORD ratings were negatively correlated with judgement error. This suggests that SA-SWORD is a reasonably precise measure of a pilot's awareness of the current location of objects relative to the operator, and the relative position of those objects over time. This suggests then that SA-SWORD has reasonable construct validity.

Vidulich and Hughes (1991) used SA-SWORD in an aircraft simulator study to compare two cockpit displays, one of which provided the pilots with additional information pertinent to their task, such as threats approaching from behind. Pilots rated their situation awareness as higher with the display that provided the additional information. This shows that SA-SWORD can discriminate between known differences in situation awareness (i.e. it has good sensitivity).

Like SART, SA-SWORD does not appear to correlate with SAGAT (Snow and Reising, 2000), although again it could well be that the two measures are assessing different concepts, both of which may ultimately be useful in understanding an operator's situation awareness (Endsley et al., 1998; Snow and Reising, 2006).

Pros

  • SA-SWORD is relatively easy to implement and can be used in simulators or in the real world.
  • SA-SWORD also has an advantage over other subjective measures (such as SART) because it relies on paired comparisons of participants' own ratings rather than absolute judgements of situation awareness. This means that problems associated with different users having different interpretations of situation awareness are reduced.

Cons

  • As with SART, SA-SWORD may not be an accurate measure of situation awareness in that people may not admit to lapses in situation awareness or may not have insight into their situation awareness.
  • SA-SWORD may be influenced by an individual's subjective performance.

SA-SWORD may be useful in providing a measure of subjective situation awareness, which may be different to the types of situation awareness assessed by other measures.

Situational Awareness Linked Indicators Adapted to Novel Tasks (SALIANT)

SALIANT was developed specifically to examine shared situation awareness. SALIANT is based on the premise that there are behavioural indicators of high and low team situation awareness (Muniz, Stout, Bowers and Salas, 1998). For example, a behavioural indicator of high team situation awareness is 'sharing of relevant information between team members'. Whereas behavioural indicators of low team situation awareness would include lack of communication, lack of listening, an argumentative crew and not noticing mistakes (Muniz et al., 1998).

Approach
There are five steps in the SALIANT method:

- Delineation of behaviours theoretically linked to team situation awareness
- Development of scenario events
- Identification of specific, observable responses
- Development of a script
- Development of an observation form (Muniz et al., 1998).

Reliability and Validity
Only a few studies have examined the reliability and validity of SALIANT, although these studies are generally positive about the method. Bowers, Weaver, Barnett and Stout (1998) examined the reliability and validity of SALIANT in a flight simulation study. Comparisons of ratings between two independent raters provided support for the reliability of this measure. To test for validity, Bowers et al. (1998) compared the SALIANT ratings with performance indicators and SAGAT. The results showed that two pairs of problem resolution dimensions of SALIANT were good predictors of performance. However, neither SAGAT nor the other dimensions of SALIANT were predictive of performance.

Pros
- This is one of the few methods to evaluate shared situation awareness.

Cons
- There is no assessment of the cognitive processes involved in team situation awareness.
- Only those behaviours that result in high team situation awareness are taken into consideration.
- Developing a suitable scenario requires extensive knowledge of tasks, standard operating procedures and regulations.
- Delineating the most effective responses is labour-intensive (Muniz et al., 1998).

Low-Event Task Subjective Situation Awareness (LETSSA)

A new technique for measuring situation awareness for low-event tasks has recently been developed and is currently being evaluated by Rose et al. (2012a; 2012b). To date, the evaluation has been based on simulator studies involving non-train drivers and experienced long-haul freight train drivers (Rose et al., 2012a; 2012b). Rather than being a measure per se, LETSSA is a technique that can be used to develop a subjective measure of situation awareness for any low-event task. Initially, a task analysis is compiled and the situation awareness requirements of the operator are highlighted. Using Endsley's (1988) model of situation awareness, questions are developed that aim to encapsulate situation awareness throughout the time being evaluated (e.g. a simulation or a section of a route).

Reliability and Validity
LETSSA is still in the very early stages of development and testing, thus there are no reliability or validity studies. Initial investigations have, however, shown that it is sensitive to manipulated changes in situation awareness (Rose et al., 2012a; 2012b).

Pros
- LETSSA is a technique that can be used for any low-event task.
- LETSSA can be used in naturalistic settings (e.g. when a train driver stops at a station or in a siding).
- LETSSA is easy to administer.
- Once a questionnaire has been developed for a task, that questionnaire can be modified for use in similar tasks, without having to begin the process over again.

Cons
- LETSSA is in its infancy and therefore validity and reliability have not been established.
- As with other subjective measures, there is a potential influence of performance.

LETSSA is a very new technique and requires full testing for reliability and validity, however, early findings are promising (Rose et al., 2012a; 2012b).

Eye-Tracking

Visual attention is a vital element of many tasks. Monitoring the external environment (e.g. outside the train cab) as well as equipment and gauges of the internal environment (e.g. within the train cab) are mostly visual activities. For this reason, some researchers posit that visual attention may be indicative of situation awareness (e.g. Chaparro, Groff, Tabor, Sifrit and Gugerty, 1999; Moore and Gugerty, 2010).

According to Kardos (2004), failures in perception account for the highest percentage of situation awareness errors. Results of tests conducted by Chaparro et al. (1999) in a motor vehicle driving simulator suggest that divided and selective attention has a significant impact on a driver's ability to maintain situation awareness. This is supported by research findings that older adults with reduced visual attention have an increased risk of being involved in an accident (Ball, Owsley, Sloane, Roenker and Bruni, 1993; Owsley et al., 1998).

Reliability and Validity

Results of investigations into eye-tracking as a measure of situation awareness have been varied. Moore and Gugerty (2010) tested the viability of eye-tracking as a measure of situation awareness, using an air traffic control task carried out by experienced air traffic controllers. They compared results from eye-tracking data with objective scores of situation awareness derived from queries made during a freeze in the simulation. They found that the percentage of time fixating on aircraft areas of interest was significantly related to situation awareness. Moore and Gugerty (2010) also used their eye-tracking data to examine the effects of focused and distributed attention and found that when attention was distributed equally across all aircraft systems, situation awareness increased, while more focused attention resulted in decreased situation awareness.

In contrast, Durso et al. (1995) did not find any significant differences between expert and novice chess players when using eye-tracking to measure situation awareness. This was despite the fact that two query-based procedures (SAGAT and SPAM) detected significant differences, as would be expected between experts and novices. In a study designed to investigate the use of eye-tracking data as a measure of individual and shared situation awareness, Hauland (2008) found mixed results, with situation awareness (as measured by distributed and focused attention) correlating with some aspects of performance but not others.

Strayer, Cooper and Drews (2004) found that, despite significant differences in object recall between a dual-task scenario (where a second task had significant attentional demands) and a single-task scenario (where the attentional demands were less), there were no differences in visual fixation on objects between the two scenarios. Specifically, participants spent the same amount of time looking at objects but did not obtain the same level of situation awareness from those looks, suggesting that the division of attention is not related to visual scanning but to deeper cognitive processing (Strayer et al., 2004).

Another problem with eye-tracking is that it is possible to be paying attention to something without looking directly at that object. Duchowski (2002) provides a good example of this, relating to astronomy, where in order to see faint stars or star clusters with the naked eye, astronomers will fixate at a point near the stars but be attending to the stars nearby.

Due to these mixed results, reliability and validity of eye-tracking as a measure of situation awareness requires further investigation and thus it should be used with caution.

Pros

- Eye-tracking is useful for learning about operators' eye fixations and visual scanning patterns but other methods, such as verbal protocol (discussed below) are required to understand their higher cognitive processes (Andersen et al., 2004).
- Eye-tracking could be useful in environments where direct measures of situation awareness are not feasible (Moore and Gugerty, 2010).

Cons

- The evidence for the effectiveness of eye-tracking measures of situation awareness is mixed.
- Eye-tracking only determines what a person is looking at and not their level of comprehension. A person's eyes can be directed at an object for a long time while thinking about something completely different (Moore and Gugerty, 2010).
- It is possible to be paying attention to something without looking directly at that object (Duchowski, 2002). Hence it is not clear that the eye-tracking measures of situation awareness have good construct validity.
- Eye-tracking is not reliable for around 10 to 20 per cent of participants (Jacob and Karn, 2003). In some cases this number is much higher. For example Moore and Gugerty (2010) were not able to collect data from five out of 16 participants owing to problems encountered with their eye-tracking equipment and laptop.

Verbal Protocol

Verbal protocols, also known as talk-aloud or think-aloud protocols, have been used extensively in research for numerous purposes, including aiding the development of computer systems, improving written instructions and as a way of generating feedback on designs of human-computer interfaces (Mehlenbacher, 1993). Verbal protocols may be conducted retrospectively (talking about cognitive processes after an event) or concurrently (talking while carrying out a task) (Ericsson and Simon, 1980).

Approach

There are two methods of concurrent verbal protocol: thinking aloud during the performance of a task (which should reveal information that the operator is currently attending to); and responding to probing questions aimed at gleaning specific information (Ericsson and Simon, 1980).

Reliability and Validity

One of the issues around the validity of the verbal protocol method is that people may not have complete access to the processes involved in creating situation awareness or if they do, the process of obtaining a verbal protocol may change the course and structure of the cognitive processes being studied (Ericsson and Simon, 1980). Thus verbal protocol may not measure what it intends to measure (construct validity) and may be obtrusive. This may be particularly true in experiments where instructions are given to participants at the beginning or where verbal reporting is required at the end of the experiment (Ericsson and Simon, 1980).

Ericsson and Simon (1980) found the following issues associated with verbal protocol:

- Concurrent verbalisations could be problematic when completing tasks where the processing of information is not verbal – performance may slow down and verbalisations may be incomplete.
- In tasks with experienced operators, automaticity (carrying out a task with subconscious cognitive processing) may cause verbalisations to be imprecise as it is difficult to explain something that is not quickly accessible in short-term memory.
- Tasks that employ complex visual processing or that have a large motor-perceptual component may result in slower performance caused by the effort required to translate actions into words.
- Effects on performance may be exacerbated if participants are asked to verbalise about specific types of information, especially if that information is not normally available for that task.
- Studies using probes to elicit specific information are more likely to affect performance than probes asking for more general information.

Pros
- Verbal protocols can be rich in data and tell the researcher a lot about the thought processes of the operator.

Cons
- Verbal protocol measures have not been extensively researched – there is a question about the extent to which they actually measure situation awareness because participants may not be able to verbalise key elements.
- Using verbal protocol to measure situation awareness does not result in a score which can be used to determine levels of situation awareness (e.g. poor, good or excellent).
- Verbal protocols are generally labour-intensive, especially with regard to analysing the data, although tools have been developed to assist with analysis of verbal protocol data (e.g. SHAPA, Sanderson, 1990).

Verbal protocol is a useful way of obtaining richer more qualitative information about situation awareness and is particularly useful as part of a battery of situation awareness measures.

Performance-Based Measures

Performance-based measures focus on the outputs of individuals rather than attempting to assess the accuracy of their mental models (Pritchett, Hansman and Johnson, 1996).

Approach

For performance-based measures, the situation is manipulated such that an action is required if the operator has sufficient situation awareness (Pritchett et al., 1996). If that action is not carried out, it is assumed to be due to the operator's lack of awareness of the situation. Performance-based measures are therefore implicit as situation awareness is implied based upon the specific information to which the individual responds (Pritchett et al., 1996).

In certain situations, performance-based measures may be more relevant or provide more important information than a knowledge-based measure such as SAGAT. Pritchett et al. (1996) argue that although both SAGAT and performance measures provide information about situation awareness, only the performance-based measure will reveal how an operator will act upon certain information. In addition, performance-based measures are able to determine how operators perceive the reliability of the knowledge they have gathered, and can determine real-time responses rather than responses that have been planned (Pritchett et al., 1996).

Three types of performance-based measures have been identified (Endsley, 1995a):

- Global measures: infer situation awareness from the overall performance of an operator.
- External task measures: involve altering or removing certain pieces of information and then measuring the time it takes for the operator to respond.
- Imbedded task measures: focus on sub-tasks, a measure which could be used to determine the influence of a display on a specific sub-task.

Reliability and Validity

Performance measures are based on the assumption that there is a direct relationship between situation awareness and performance, that is, good situation awareness leads to good performance (Sarter and Woods, 1995; Uhlarik and Comerford, 2002). This assumption is problematic because it is possible to have good situation awareness and still perform poorly and it is possible to have poor situation awareness and still perform well (Salmon et al., 2006). For example, a train driver may pass a yellow signal without perceiving or comprehending it but by the time the next signal is sighted, it is green (upgraded from red) and so performance will be unaffected by the lack of awareness of the yellow signal. Thus the validity of performance measures is questionable.

Sarter and Woods (1995) suggest that debriefings conducted to determine why particular behaviours did or did not occur will help to address the problem of the assumed relationship between situation awareness and performance. Another approach is to use verbal protocol measures of situation awareness (see 'Verbal Protocol' above) in conjunction with performance measures.

Pros
- Performance-based measures are relatively easy to use, objective and usually non-intrusive.
- When using simulators for research, computer programs can be used to automatically record the data, making data collection relatively easy (Endsley, 1995a).

Cons
- Global measures do not provide diagnostic information because they do not provide adequate information to be able to determine the causes of poor performance.
- There are numerous factors other than situation awareness that may influence performance.
- External task measures are intrusive as the operator is required to investigate what happened while still maintaining performance on their primary task, which may lead to misleading results and change the ongoing task.
- Imbedded task measures only provide information on the operator's situation awareness for the element of interest and do not take into consideration the potential impact upon other elements. It may be that an improvement in situation awareness for one element of a task may reduce situation awareness on another element and thus results will be misleading (Endsley, 1995b).

Performance measures provide a slightly different view of situation awareness and may be useful in conjunction with other measures of situation awareness. They can be used for testing systems to see if sufficient situation awareness is being generated to allow individuals to perform the correct actions (Pritchett et al., 1996).

Observer Ratings

As the name suggests, observer ratings are elicited from observers, usually subject matter experts, who observe operators and/or listen to operators' verbalisations in order to assess their situation awareness. Some of the methods considered above can also be employed by observers. Observer ratings are most commonly used in naturalistic settings because they are non-intrusive (Salmon et al., 2009). Behaviours considered relevant to situation awareness are pre-defined and observers look for these behaviours in order to give operators a situation awareness rating (Salmon et al., 2009).

Reliability and Validity
Concerns have been raised regarding the validity of observer ratings due to uncertainty of the ability of observers to rate another individual's situation

awareness and doubt as to whether overt performance is a true reflection of an operator's situation awareness (Salmon et al., 2009).

Endsley (1995a) argues that subjective observer ratings are of limited value because judgements can only be made by observing the behaviour of the operator and/or listening to verbalisations made during the course of the task. This may provide information useful for diagnostic purposes but would be incomplete as an operator is not likely to verbalise every thought relating to situation awareness (Endsley, 1995a). An observer cannot be sure whether the omission of an important variable is due to lack of situation awareness or simply that the operator has not mentioned it (Endsley, 1995a).

Pros
- Observer ratings are non-intrusive and thus have no impact upon the task being performed (Salmon et al., 2009).
- Due to the non-intrusive nature of observer ratings, they can be used in naturalistic settings (Salmon et al., 2009).

Cons
- Observer ratings can only measure situation awareness of a participant by observable behaviours (Endsley, 1995a).
- Observer ratings can be affected by bias (Stanton et al., 2005).
- Being observed may also cause participants to change their behaviour (Stanton et al., 2005).
- It may be difficult to organise a sufficient number of observers to participate in a study (Salmon et al., 2010).

Summary

There are a large number of methods for assessing situation awareness, each with its own strengths and limitations. Table 10.1 below presents each measurement tool and technique with a summary of its validity/reliability, data collection format (simulator vs on-the-job assessment) and the method of data collection.

This table allows the operator to do a quick comparison of all methods to support judgements on the most appropriate method of situation awareness measurement for their operating environment, based on the availability of simulators and resources.

There are no hard and fast rules about which method to choose, since each contributes to a slightly different understanding of situation awareness. However, due to the limitations of all measures of situation awareness, Salmon et al. (2006) suggest a multiple measure approach where a variety of different methods are used.

Table 10.1   Summary of situation awareness measurement tools and techniques

Measure/Tool	Validity/ Reliability	Format	Data collection method
SAGAT	High	Simulator-based assessment	Researcher-led question and answer
SPAM	Medium	Simulator-based assessment On-the-job assessment	Researcher-led question and answer
SART – 3-D	Medium – Low	Simulator-based assessment On-the-job assessment	Researcher-led question and answer
SART – 10-D	High	Simulator-based assessment On-the-job assessment	Researcher-led question and answer
SA-SWORD	Medium	Simulator-based assessment On-the-job assessment	Researcher-led question and answer
SALIANT	Unknown	Simulator-based assessment	Researcher observation
LETSSA	Unknown	Simulator-based assessment On-the-job assessment	Researcher-led question and answer
Eye-tracking	Medium – Low	Simulator-based assessment On-the-job assessment	Data analysis *Recommend simultaneous verbal protocol*
Verbal protocol	Medium – Low	Simulator-based assessment On-the-job assessment	User verbalisation of tasks with researcher observation Researcher-led question and answer
Performance- based	Medium – Low	Simulator-based assessment	Researcher observation *Recommend simultaneous verbal protocol*
Observer ratings	Low	Simulator-based assessment On-the-job assessment	Researcher observation *Recommend simultaneous verbal protocol*

Part 2 Conclusions

In this part of the chapter we have considered the different methods that are available to measure situation awareness. This review provides information that should be considered when deciding which measure/s to use to assess the impact of a new technology on both individuals and teams.

CHAPTER CONCLUSION

Situation awareness is a key concept that underpins the complex actions performed by train drivers and network controllers. In essence, situation awareness describes the way that people build and maintain an ongoing understanding of the environment in which they are operating. Decisions about the control of the train or train control system are underpinned by that operator's situation awareness. When introducing a new technology it is important to examine the effects that this will have on the individual, the people with whom that individual connects and the broader social network in which work is carried out. To provide a more sophisticated understanding of situation awareness a number of different ways in which researchers have conceptualised individual and team situation awareness were discussed in the first part of the chapter. The important topic of how to measure situation awareness was discussed in the second half and a number of different measures were considered together with an analysis of the strengths and limitations of each. To provide the most effective evaluation of situation awareness it is argued that a number of different methods should be employed. As technological systems become increasingly more advanced it is important to consider the impact that a technology has on the way that both the operator and others build and maintain an understanding of both the task and the surrounding technological, physical and social environment in which rail operations are conducted.

Acknowledgements

The authors are grateful to the CRC for Rail Innovation (established and supported under the Australian Government's Cooperative Research Centres program) for the funding of this research. We are also grateful to all those who participated in our simulator study, and to the rail organisations that provided assistance in recruiting experienced train drivers. Thanks also to Dr Vincent Rose for his assistance in redrawing many of the figures in this chapter.

References

Adams, M.J., Tenney, Y.J., and Pew, R.W. (1995). Situation awareness and the cognitive management of complex systems. *Human Factors, 37*(1), 85–104.

Andersen, H.H.K., Andersen, H.B., Hilburn, B.G., Zon, R., Blechko, A., Ober, J., and Hauland, G. (2004). *Coding and inferences from visual and other behavioural data* [Report No. Riso-R-1474(EN)]. Roskilde, Denmark: Riso National Laboratory.

Artman, H., and Garbis, C. (1998). Situation Awareness as Distributed Cognition. In T. Green, L. Bannon, C. Warren and J. Buckley (eds.), *Proceedings of the 9th Conference of Cognitive Ergonomics: Cognition and cooperation* (pp. 151–6). Limerick, Ireland.

Australian Transport Safety Bureau. (2005). *Derailment of Cairns Tilt Train VCQ5, 15 November 2004* (Rail Occurrence Investigation 2004/007). Canberra: Australian Transport Safety Bureau.

Ball, K., Owsley, C., Sloane, M.E., Roenker, D.L., and Bruni, J.R. (1993). Visual attention problems as a predictor of vehicle crashes in older drivers. *Investigative Ophthalmology & Visual Science, 34*(11), 3110–23.

Bedny, G., and Meister, D. (1999). Theory of activity and situation awareness. *International Journal of Cognitive Ergonomics, 3*(1), 63–72.

Bolstad, C.A., and Endsley, M.R. (2000). The Effect of Task Load and Shared Displays on Team Situation Awareness. *Proceedings of the Human Factors and Ergonomics Society Annual Meeting, 44*(1), 189–92.

Bolton, M.L., and Bass, E.J. (2009). Comparing perceptual judgment and subjective measures of spatial awareness. *Applied Ergonomics, 40*(4), 597–607.

Bowers, C., Weaver, J., Barnett, J., and Stout, R. (1998). Empirical Validation of the SALIANT Methodology. *RTO HFM Symposium on Collaborative Crew Performance in Complex Operational Systems* (pp. 12-1–12-6), *20–22 April 1998, Edinburgh, Scotland.*

Chaparro, A., Groff, L., Tabor, K., Sifrit, K., and Gugerty, L.J. (1999). Maintaining Situational Awareness: The Role of Visual Attention. *Proceedings of the Human Factors and Ergonomics Society Annual Meeting, 43*(23), 1343–7.

Duchowski, A.T. (2002). A breadth-first survey of eye tracking applications. *Behavior Research Methods: Instruments, and Computers, 34*(4), 455–70.

Durso, F.T., Bleckley, M.K., and Dattel, A.R. (2006). Does situation awareness add to the validity of cognitive tests? *Human Factors, 48*, 721–33.

Durso, F.T., and Dattel, A.R. (2004). SPAM: the real-time assessment of SA. In S. Banbury and S. Tremblay (eds.), *A Cognitive Approach to Situation Awareness: Theory and Application* (pp. 137–54). Aldershot: Ashgate Publishing.

Durso, F.T., and Dattel, A.R. (2006). Expertise and transportation. In K. Anders Ericsson, N. Charness, P.J. Feltovich and R.R. Hoffman (eds.), *The Cambridge Handbook of Expertise and Expert Performance* (pp. 355–71). Cambridge: Cambridge University Press.

Durso, F.T. Hackworth, C.A., Truitt, T.R., Crutchfield, J., Nikolic, D., and Manning, C.A. (1999). *Situation Awareness As a Predictor of Performance in En Route Air Traffic Controllers* (Report No. DOT/FAA/AM-99/3). Washington, DC: Office of Aviation Medicine, Federal Aviation Administration, US Department of Transportation.

Durso, F.T., Truitt, T.R., Hackworth, C.A., Crutchfield, J.M., Nikolic, D., Moertl, P.M. ... Manning, C.A. (1995). Expertise and chess: a pilot study comparing situation awareness methodologies. In D.J. Garland and M.R. Endsley (eds.), *Experimental Analysis and Measurement of Situation Awareness* (pp. 295–304). Daytona Beach, FL: Embry-Riddle Aeronautical Press.

Endsley, M.R. (1988). Situation Awareness Global Assessment Technique (SAGAT). *Proceedings of the National Aerospace and Electronics Conference* (pp. 789–95). New York: IEEE.

Endsley, M.R. (1990). Predictive Utility of an Objective Measure of Situation Awareness. *Proceedings of the Human Factors and Ergonomics Society Annual Meeting, 34*(1), 41–5.

Endsley, M.R. (1993). Situation Awareness and Workload: Flip Sides of the Same Coin. *Proceedings of the 7th International Symposium on Aviation Psychology* (pp. 906–11). Columbus, OH: Ohio State University, Department of Aviation.

Endsley, M.R. (1995a). Measurement of situation awareness in dynamic systems. *Human Factors, 37*(1), 65–84.

Endsley, M.R. (1995b). Toward a theory of situation awareness in dynamic systems. *Human Factors, 37*(1), 32–64.

Endsley, M.R. (2000). Direct measurement of situation awareness: validity and use of SAGAT. In M.R. Endsley and D.J. Garland (eds.), *Situation Awareness Analysis and Measurement* (pp. 147–73). Mahwah, NJ: Lawrence Erlbaum Associates.

Endsley, M.R., and Bolstad, C.A. (1994). Individual differences in pilot situation awareness. *The International Journal of Aviation Psychology, 4*(3), 241–64.

Endsley, M.R., Selcon, S.J., Hardiman, T.D., and Croft, D.G. (1998). A Comparative Analysis of SAGAT and SART for Evaluations of Situation Awareness. *Proceedings of the Human Factors and Ergonomics Society Annual Meeting, 42*(1), 82–6.

Ericsson, K.A., and Simon, H.A. (1980). Verbal reports as data. *Psychological Review, 87*(3), 215–50.

Fracker, M., (1991). *Measures of Situation Awareness: Review and Future Directions* (Report No. AL-TR-1991-0128). Wright-Patterson Air Force Base: Armstrong Laboratories.

Gugerty, L.J. (1997). Situation awareness during driving: explicit and implicit knowledge in dynamic spatial memory. *Journal of Experimental Psychology: Applied, 3*, 42–66.

Hauland, G., (2008). Measuring individual and team situation awareness during planning tasks in training of en route air traffic control. *The International Journal of Aviation Psychology, 18*(3), 290–304.

Hourizi, R., and Johnson, P. (2003). Towards an explanatory, predictive account of awareness. *Computers & Graphics, 27*, 859–72.

Hunn, B.P. (2001). Definition-based situation awareness. *Proceedings of the Human Factors and Ergonomics Society Annual Meeting, 45*(25), 1777–9.

Jacob, R.J.K., and Karn, K.S. (2003). Eye tracking in human-computer interaction and usability research: ready to deliver the promises. In R. Radach, J. Hyona and H. Deubel (eds.), *The Mind's Eye: Cognitive and Applied Aspects of Eye Movement Research* (pp. 573–605). Oxford: Elsevier Science BV.

Kardos, M. (2004). *Automation, Information Sharing and Shared Situation Awareness* (DSTO-GD-0400). Edinburgh, South Australia: DSTO Systems Sciences Laboratory.

Luther, R., Livingstone, H., Gipson, T., and Grimes, E. (2005, November). Methodologies for the Application of Non-Rail Specific Knowledge to the Rail Industry. *Second European Conference on Rail Human Factors, York.*

Ma, R., and Kaber, D.B. (2007). Situation awareness and driving performance in a simulated navigation task. *Ergonomics, 50*(8), 1351–64.

Mehlenbacher, B. (1993). Software Usability: Choosing Appropriate Methods for Evaluating Online Systems and Documentation. *Proceedings of the 11th Annual International Conference on Systems Documentation, 5–8 October 1993, Kitchener, ON, Canada.*

Mogford, R.H. (1997). Mental models and situation awareness in air traffic control. *The International Journal of Aviation Psychology, 7*(4), 331–41.

Moore, K., and Gugerty, L. (2010). Development of a Novel Measure of Situation Awareness: The Case for Eye Movement Analysis. *Proceedings of the Human Factors and Ergonomics Society Annual Meeting, 54*(19), 1650–54.

Muniz, E.J., Stout, R.J., Bowers, C.A., and Salas, E. (1998). A methodology for measuring team situational awareness: Situational Awareness Linked Indicators Adapted to Novel Tasks. *RTO HFM Symposium on Collaborative Crew Performance in Complex Operational Systems* (pp. 11-1–11-8). Neuilly-Sur-Seine Cedex, France: NATO Research and Technology Organization.

Neisser, U. (1976). *Cognition and Reality: Principles and Implications of Cognitive Psychology.* San Francisco: Freeman.

Orasanu, J.M. (1995). Situation Awareness: Its Role in Flight Crew Decision Making. In R.S. Jensen and L.A. Rakovan (eds.), *Proceedings of the Eighth International Symposium on Aviation Psychology, April 24–27, 1995* (pp. 734–9). Columbus, OH: Ohio State University, Dept. of Aerospace Engineering, Applied Mechanics and Aviation.

Owsley, C., Ball, K., McGwin, G., Sloane, M.E., Roenker, D.L., White, M.F., and Overley, E.T. (1998). Visual processing impairment and risk of motor vehicle crash among older adults. *The Journal of the American Medical Association, 279*(14), 1083–8.

Pritchett, A.R., Hansman, R.J., and Johnson, E.N. (November 1996). Use of Testable Responses for Performance-based Measurement of Situation Awareness. *International Conference on Experimental Analysis of Measurement of Situation Awareness.* Daytona Beach, FL.

Rose, J.A., Bearman, C., and Dorrian, J. (2012a). An evaluation of the Low-Event Task Subjective Situation Awareness (LETSSA) Technique. In N.A. Stanton (ed.), *Advances in Human Aspects of Road and Rail Transportation* (pp. 690–703). Abingdon, Oxon: Taylor & Francis, CRC Press.

Rose, J.A., Bearman, C., and Dorrian, J. (2012b). Constructing and Evaluating the Low-Event Task Subjective Situation Awareness (LETSSA) Measure. *10th International Symposium of the Australian Aviation Psychology Association, 19–22 November 2012, Manly, NSW, Australia.*

Roth, E., and Multer, J. (2009). *Technology Implications of a Cognitive Task Analysis for Locomotive Engineers* (Report No. DPT/FRA/ORD-09/03). Washington, DC: US Department of Transportation, Federal Railroad Administration.

Salas, E., Prince, C., Baker, D.P., and Shrestha, L. (1995). Situation awareness in team performance: implications for measurement and training. *Human Factors, 37*(1), 123–36.

Salmon, P., Stanton, N., Walker, G., and Green, D. 2006. Situation awareness measurement: A review of applicability for C4i environments. *Applied Ergonomics, 37*(2), 225–38.

Salmon, P.M., Stanton, N.A., Walker, G.H., Baber, C., Jenkins, D.P., McMaster, R., and Young, M.S. (2008). What really is going on? Review of situation awareness models for invididuals and teams. *Theoretical Issues in Ergonomics Science, 9*(4), 297–323.

Salmon, P.M., Stanton, N.A., Walker, G.H., Jenkins, D., Ladva, D., Rafferty, L., and Young, M. (2009). Measuring situation awareness in complex systems: Comparison of measures study. *International Journal of Industrial Ergonomics, 39*(3), 490–500.

Salmon, P.M., Stanton, N.A., Walker, G.H., Jenkins, D.P. and Rafferty, L. (2010). Is it really better to share? Distributed situation awareness and its implications for collaborative system design. *Theoretical Issues in Ergonomics Science, 11*(1–2), 58–83.

Sanderson, P.M. (1990, October). Verbal Protocol Analysis in three Experimental Domains Using Shapa. *Proceedings of the Human Factors and Ergonomics Society Annual Meeting, 34*(17), 1280–84.

Sarter, N.B., and Woods, D.D. (1991). Situation awareness: A critical but ill-defined phenomenon. *The International Journal of Aviation Psychology, 1*(1), 45–57.

Sarter, N.B., and Woods, D.D. (1995). How in the world did we ever get into that mode? Mode error and awareness in supervisory control. *Human Factors, 37*(1), 5–19.

Selcon, S.J., and Taylor, R.M. (1990). Evaluation of the situational awareness rating technique (SART) as a tool for aircrew systems design. *Situational Awareness in Aerospace Operations* (AGARD-CP-478). Neuilly-Sur-Seine, France: NATO-Advisory Group for Aerospace Research and Development.

Selcon, S.J., Taylor, R.M., and Koritsas, E. (1991). Workload or Situational Awareness? TLX vs. SART for Aerospace Systems Design Evaluation. *Proceedings of the Human Factors Society Annual Meeting, 35*(2), 62–6.

Smith, K., and Hancock, P.A. (1995a). The Risk Space Representation of Commercial Airspace. In R.S. Jensen and L.A. Rakovan (eds.), *Proceedings of the Eighth International Symposium on Aviation Psychology, April 24–27,*

1995. Columbus: Ohio State University, Dept. of Aerospace Engineering, Applied Mechanics and Aviation.

Smith, K. and Hancock, P.A. (1995b). Situation awareness is adaptive, externally directed consciousness. *Human Factors, 37*(1), 137–48.

Snow, M.P., and Reising, J.M. (2000). Comparison of Two Situation Awareness Metrics: SAGAT and SA-SWORD. *Proceedings of the Human Factors and Ergonomics Society Annual Meeting, 44*(13), 49–52.

Stanton, N.A., Salmon, P.M., Walker, G.H., Baber, C., and Jenkins, D.P. (2005). *Human Factors Methods: A Practical Guide for Engineering and Design.* Aldershot: Ashgate Publishing.

Stanton, N.A., Stewart, R., Harris, D., Houghton, R.J., Baber, C., McMaster, R., … Green, D. (2006). Distributed situation awareness in dynamic systems: theoretical development and application of an ergonomics methodology. *Ergonomics, 49*(12–13), 1288–311.

Strayer, D.L., Cooper, J.M., and Drews, F.A. (2004). What do Drivers Fail to See When Conversing on a Cell Phone? *Proceedings of the Human Factors and Ergonomics Society Annual Meeting, 48*(19), 2213–17.

Taylor, R.M. (1990). Situational Awareness Rating Technique (SART): The development of a tool for aircrew systems design. *Situational Awareness in Aerospace Operations* (AGARD-CP-478). Neuilly Sur Seine, France: NATO-AGARD.

Uhlarik, J., and Comerford, D.A. (2002). *A Review of Situation Awareness Literature Relevant to Pilot Surveillance Functions* (DOT/FAA/AM-02/3). Washington, DC: Federal Aviation Administration, US Department of Transportation.

Vidulich, M.A., and Hughes, E.R. (1991). Testing a subjective metric of situation awareness. *Proceedings of the Human Factors Society Annual General Meeting,* 35(18), 1307–11.

Index